Environmental Sustainability

Environmental Sustainability

Water and Waste Management Policy in the European Union and the Czech Republic

Gamze Tanil

Foreword by Bedřich Moldan

LEXINGTON BOOKS
Lanham • Boulder • New York • London

Published by Lexington Books
An imprint of The Rowman & Littlefield Publishing Group, Inc.
4501 Forbes Boulevard, Suite 200, Lanham, Maryland 20706
www.rowman.com

6 Tinworth Street, London SE11 5AL, United Kingdom

British Library Cataloguing in Publication Information Available

Library of Congress Cataloging-in-Publication Data

Library of Congress Control Number: 2020949052
ISBN: 978-1-7936-3387-3 (cloth)
ISBN: 978-1-7936-3388-0 (electronic)

Contents

List of Figures and Tables

FIGURES

TABLES

Foreword

In November 1989 the Czech Republic experienced the so-called Velvet Revolution that brought along a new democratic regime. Environmental aspects represented an extremely important element of the transition. One must be aware of the fact that during the time of the totalitarian regime in then Czechoslovakia the environmental situation was extremely bad both nationally and in some more affected localities and regions in particular. Moreover, there even did not exist any adequate national system of environmental protection; for example, a Ministry of the Environment or similar institution was nonexistent. Even if the regime did not allow informing the public realistically about the important aspects of pollution of air, water, and other parts of the environment people were well aware of the acute dangers connected with the high levels of pollution and harm of their health and indeed lives.

So it was very logical that people asked the new government to apply the appropriate measures quickly and ensure a rapid and thorough improvement. The new legislation was indeed enacted very soon along with important institutions starting with the Ministry of the Environment, State Fund for the Environment, Czech Environmental Inspectorate, and others. All the measures enjoyed a broad public support. And the factual improvement was quite visible during only a few years.

In 1999 the negotiation on accession to the European Union started, and it was rather smoothly finished two years later. This process has not changed fundamentally the already existing rather successful system of environmental protection even if certainly many novel elements were added. In the time of entering the EU in 2004 we may say that the whole process was fully finished with basically no transition period. Since then, the further evolution of environmental legislation and other measures followed that of the EU.

Gamze Tanil focused on two important areas of the environment: water and waste policy. She analyzed very carefully the whole series of legal acts and other documents together with the evolution of the factual situation. She presents a realistic and detailed account how the policies and state of the two important areas evolved. The example of the Czech Republic as a country undergoing the transition from the state of rather catastrophic environmental pollution to the state that may be called a European normal is well documented. I am sure that her contribution is worth reading for people interested in the evolution of environmental protection in general and in the given areas in particular.

Bedřich Moldan, the first Minister of the Environment of the Czech Republic.

Acknowledgments

The world is a better place thanks to people who want to develop and lead others. To all individuals who strive to grow and help others grow, I want to say thank you for being the inspiration and foundation of my contribution to the literature on environmental sustainability.

I wish to offer my special thanks to the University of West Bohemia. This book would not have been completed without their support. I would like to express my very great appreciation to the Dean of the Faculty of Philosophy, Prof. Dr. David Šanc, for his valuable and constructive suggestions during the planning and development of this research work. His willingness to give his time so generously has been very much appreciated. My special thanks are extended to my peers and colleagues at the Department of Politics and International Relations for their cooperation and support. Petr, Martina, Přemysl, Jiří, Šárka, and Pavel, your friendship and encouragement during this research period were greatly appreciated.

I sincerely thank Dr. Magnus Lindh for his useful critiques of this research work. The advice given by him has been a great help in providing a new perspective. Last but not least, I would like to express my gratitude to the respondents of my interviews for their willingness to spend time to answer my questions.

Also, thanks to everyone on the Lexington team who helped me so much. Special thanks to Joseph, the ever-patient Publishing Manager.

Finally, I wish to thank my family and friends for their support and encouragement throughout my research.

Chapter 1

Introduction

THEORETICAL BACKGROUND: HOW TO STUDY EUROPEANIZATION?

This section explains how the concept of "Europeanization" can be used to investigate the effects that the EU policies may have on domestic public policy in the member states.

An analysis of academic literature on Europeanization should start by a conceptual analysis since this is a fundamental step in political science. What does the popular concept "Europeanization" refer to? What is "Europeanized" and how is it "Europeanized"? Different authors assign different meanings to this concept, so I will briefly discuss this literature here.

A review of the literature demonstrates two different approaches to Europeanization originating from policy analysis and international relations, which have different core interests and should therefore be kept separate. Policy analysis is concerned with the "varying impact of European integration on domestic arrangements and structures."[1] In this approach, the level of analysis is the domestic system, and the main analytical construct is exploring the impact of the EU. Several political scientists conceptualized the domestic effect of Europeanization as a process of change at the domestic level in which the member states adapt their processes, policies, and institutions to new practices, norms, rules, and procedures that emanate from the emerging European system of governance.[2] Thus, this will be the approach to be used throughout this book; that is to say, I conceptualize Europeanization as "a top-down process impacting on the member states."[3] This choice is also related to the policy area under consideration. This top-down approach can be justified based on the fact that it is better suited to a regulatory policy area such as the environment.

The most precise and useful definition of Europeanization was put forward by Claudio Radaelli, who stated: "Europeanization refers to processes of (a) construction (b) diffusion and (c) institutionalization of formal and informal rules, procedures, policy paradigms, styles, ways of doing things and shared beliefs and norms which are first defined and consolidated in the EU policy process and then incorporated in the logic of domestic discourse, identities, political structures and public policies."[4] In applying this concept to the domestic policy analysis, this book investigates (c), looking at how the outcomes of EU policy processes were institutionalized in the member states. So, the main subject of this book is the public policy, not political structures, structures of representation, or cognitive and normative structures. Study of public policies embraces the study of content of policies (objectives and policy instruments), the legal and administrative structures that have been established to oversee them, and the dominant style in which policy is made and implemented. So, the approaches originating from a political perspective are neglected here.

Ladrech puts emphasis on Europeanization as a process. He argues that "Europeanization is an incremental process reorienting the direction and shape of politics to the degree that EU political and economic dynamics become part of the organizational logic of national politics and policy-making."[5] Changes in organizational logic here refer to "the adaptive processes of organizations to a changed or changing environment."[6] Thus, he concludes, "organizations respond to changes in the perceptions of interest and value that occur in the principles, norms and institutional design of the regime in which they are embedded."[7]

Other political scientists also conceptualized the domestic effect of Europeanization as a process of change at the domestic level in which the member states adapt their processes, policies, and institutions to new practices, norms, rules, and procedures that emanate from the emerging European system of governance.[8] Europeanization is most often placed within some type of institutional perspective. This view sees Europeanization as political institutionalization. This involves "the development of formal and informal rules, procedures, norms, and practices governing politics at the European, national, and subnational levels."[9]

Among these more or less similar explanations of Europeanization, the approach to be used throughout this book is the one which conceptualizes Europeanization as "a top down process impacting on the member states."[10] This choice is also related with the policy area under consideration. This top-down approach can be justified based on the fact that it is better suited to a regulatory policy area such as environment.

This literature review revolves around a series of questions. What is policy transfer? Why is there policy transfer? What is transferred? What are the mechanisms of policy transfer? Are there different degrees of policy transfer? What problems are there with the existing literature?

This discussion starts with the ideal conditions for domestic change. The concept of "goodness of fit" is used in several studies of Europeanization.[11] Most academic studies in this field share the idea that "a certain amount of misfit between European-level and domestic processes, policies, or institutions constitutes the necessary condition for expecting any change."[12] The proposition here is that in order to create pressure for adaptation on the member state there must be some misfit or incompatibility between European and domestic policies, processes, and institutions: "the lower the compatibility between European and domestic processes, policies, and institutions, the higher the adaptational pressure."[13] The logic here is that if Europeanization fits perfectly well with domestic ways of doing things, there is no need for domestic change. For example, if European environmental regulations match with domestic policies, member states do not need to change their legal provisions. By confirming this proposition, Knill and Lehmkuhl argue that "variation in domestic change can be primarily explained by varying levels of European adaptation pressure, with non-adaptation being the expected scenario in cases of either complete fit or fundamental misfit, while adaptation is to be expected in cases where European policies require merely incremental changes in existing domestic arrangements."[14]

The form of misfit can be either policy misfit or institutional misfit. In the former, domestic policy may not fit the European rules and regulations, thus European policy challenges national policy goals, standards, instruments, or techniques used to achieve policy goals. In the latter, Europeanization may challenge the roles of central and territorial institutions in implementing domestic rules and procedures, and may result in creation of new ones.

As indicated by this brief review of literature, "an institutional explanation based on the compatibility of European and domestic arrangements provide a promising and useful approach in accounting for varying patterns of domestic adjustment in the light of European requirements."[15] For this reason, the concept of goodness of fit is employed in this book and it is tested within the environmental policy.

When there is goodness of fit between the domestic and the European structures in a policy area, then we expect the European policy to be transferred to the domestic structure easily. Policy transfer is defined as a "process in which knowledge about policies, administrative arrangements, institutions, and ideas in one political setting is used in the development of policies, administrative arrangements, institutions, and ideas in another political setting."[16] In this context, the quasi-federal system of the European Union is conceptualized as "a massive policy transfer platform from dominant countries and/or winning advocacy coalitions to other countries and coalitions."[17]

Thus, policy transfer is one of the principal mechanisms of Europeanization, and it is one of the processes at the heart of the case studies.

The EU-accession process produces two varieties of policy transfer: hard transfers of programs and implementation, and soft transfers of ideas, concepts, and attitudes. Hard transfer emerges when the EU transfers its rules, procedures, and policy paradigms to member states. Soft transfer emerges when the EU transfers its policy styles, ways of doing things, and shared beliefs and norms to the member states.

Within the framework of policy transfer, Dolowitz and Marsh identify seven objects of transfer between states: "policy goals, structure and content; policy instruments or administrative techniques; institutions; ideology; ideas and attitudes; and negative lessons."[18] Similarly, Bennett explains policy transfer as "convergence of policy goals, convergence of policy content, convergence of policy instruments, and convergence of policy outcomes or consequences."[19] Convergence of *policy goals* signifies a coming together of intent to deal with common policy problems. *Policy content* is defined as the more formal manifestations of government policy, such as statutes, administrative regulations, court decisions, and so on. Convergence on *policy instruments* investigates the institutional tools available to administer policy, whether regulatory, administrative, or judicial. *Policy outcomes* refer to the results of implementation (either positive/negative or effective/ineffective).[20] Based on this approach, this book analyzes the policy transfer within the framework of policy goals, policy content, policy instruments, and policy outcomes.

When researchers seek for the mechanisms of policy transfer from the supranational level to the domestic level, they find no unique explanation for their question. Different authors consider different explanatory factors in order to account for the change and persistence of domestic institutions in the light of European requirements. For example, Knill and Lehmkuhl identify three ways in which the EU can exert influence on member states as positive integration, negative integration, and framing integration.[21] *Positive integration* is found "when European policy-making triggers domestic change by prescribing concrete institutional requirements with which member states must comply; that is, when EU directives and regulations prescribe positively an institutional model to which domestic arrangements have to be adjusted, with limited national discretion."[22] In this mechanism, EU policies are explicitly directed at replacing existing domestic regulatory arrangements. The examples of this type of policy transfer are the EU policy on environmental protection, health and safety at work, and consumer protection. On the other hand, *negative integration* occurs where "EU legislation alters the domestic rules of the game, and hence the distribution of power and resources between domestic actors."[23] However, direct institutional impact of negative integration is rather limited; member states are not directly forced to replace their regulatory arrangements in favor of an institutional model prescribed by European legislation. The single market is an example of this type of policy transfer. *Framing integration* is the weakest mechanism "where European policy neither prescribes concrete institutional requirements nor modifies the institutional context for strategic interaction, but affects domestic arrangements indirectly by altering the beliefs and expectations of domestic actors."[24] This mechanism is based on a cognitive logic.

From another point of view, DiMaggio and Powell put forward three forms of policy transfer mechanisms: coercive processes, mimetic processes, and normative pressures.[25] *Coercive processes* result from "both formal and informal pressures exerted on organizations by other organizations upon which they are dependent."[26] Since they stem from political influence, as the dependency increases, similarity of structure and behavior increases as well. *Mimetic processes* operate "when goals are ambiguous, or when the environment creates symbolic uncertainty."[27] Imitation might then be a response to uncertainty since it has the advantage that it provides an easy and cheap way of finding quick solutions to immediate problems. The modeled organization may be unaware of the modeling or may have no desire to be copied; it merely serves as a convenient source of practices that the borrowing organization may use. *Normative pressures* work when credentials, professional status, and ethical concerns are at issue. It is the collective struggle of members of an occupation to define the conditions and methods of their work, and to establish a cognitive base and legitimization for their occupational autonomy.

Bennett employs another conceptualization to explain the processes of policy transfer: emulation, elite networking, harmonization, and penetration.[28] For Bennett, *emulation* is characterized by "the utilization of evidence about a program or programs from overseas and drawing of lessons from that experience."[29] This process of lesson-drawing produces a convergence on some dimension of policy effort. It arises from the sense of collective insecurity under conditions of policy innovation and may explain the convergence of policy goals, of policy content, or of policy instruments. Secondly, *elite networking* is characterized by "the existence of shared ideas amongst a relatively coherent and enduring network of elites engaging in regular interaction at the transnational level."[30] Convergence in this sense is the result of an identifiable elite bound by knowledge and expertise of a common policy problem and a shared concern for its resolution. *Harmonization* is characterized by "the recognition and resolution of a common problem through the structures and processes of an international regime and is driven by a recognition of interdependence, a vague notion signifying a reliance on others for the performance of specific tasks to ensure complete and successful implementation."[31] The increasing transnationalization of many policy problems requires intergovernmental and supranational institutions to facilitate common responses to common problems in order to mitigate the unintended external consequences of domestic policy. Environmental policy is a good example of this type of policy area, where harmonization takes place. So, harmonization emerges from the perceived need to cooperate in order to achieve desired results in a policy area. Finally, *penetration* emerges "when states are forced to conform to actions taken elsewhere by external actors."[32] Most evidence of convergence through penetration lies in the national business in successfully securing a common regulatory framework for its products. Under certain conditions, different multinational corporations may recognize a common interest in the harmonization of policy and in achieving a substantial uniformity of national policies.

These different mechanisms of policy transfer are not excluding, rather they may work simultaneously or dominate different phases of the process of adaptational change. It is quite plausible that, when policy convergence is analyzed over time, different processes can be observed to be at work. So, if the policy analysis is cast within a temporal and comparative framework, different processes may be observed intermingling with each other, with some more relevant at some times and for some dimensions of policy. Thus, the conclusion is that policy convergence is not a once-and-for-all process; rather it should be conceptualized in dynamic terms.

It should also be noted that this research does not attempt to clarify which mechanism(s) of policy transfer are at play in the process of Europeanization of the environmental policy in the Czech Republic. The main reason for this decision is the lack of an empirical tool to distinguish these mechanisms and vagueness of such an analysis in a national context. As mentioned earlier, all these mechanisms may work simultaneously or dominate different phases of the process of adaptational change; thus it is not possible to identify one mechanism over another and hold it accountable for the domestic policy change.

This book analyzes the transfer of environmental policy from the European Union into the Czech Republic within the framework of policy goals, policy content, policy instruments, and policy outcomes. To assess these indicators of policy transfer, I focus on the transposition and institutionalization of the EU's water and waste management policies in the Czech Republic. The comprehensive legislative development as well as the broad variety of their regulatory requirements rendered these policy areas particularly appropriate for the investigation of domestic impacts of the EU policies. Methodologically, this book analyzes domestic systemic-level change as a dependent variable and measures it by analyzing the shifts in the policy norms and rules. The time frame covers the period from the Czech Republic's EU-membership to this day to reveal the full pattern of domestic policy change, or more precisely, to reveal how far the policy norms and rules adopted by the EU are refracted by the national structure.

Having clarified the approach, how can we develop an empirical measurement of the instances of policy transfer? Can we seek for a precise analytical method? We need to find answers to these analytical questions by focusing on the implementation phase of the policy transfer. Certainly, the EU actors (particularly the Commission) put increasing stress on the effective implementation of EU rules, not just writing them into national law. In this problematique, the issue is more what happens after policies are transferred from the EU to member states. Large-scale transfer of the EU rules into national law in member states certainly occurs, but what happens in the phase of policy implementation? We need to ask if it is only the political rhetoric that was Europeanized or if there are clear results of the policy implementation (either positive/negative or effective/ineffective). Thus, it is the implementation stage which is critical to understand how the EU affects domestic change in the member states and explain if there is completeness of policy transfer.

ANALYTICAL FRAMEWORK

Having examined the broad concept of Europeanization, I now turn to look at the empirical analysis of Europeanization of national public policy. To provide an empirical tool to measure Europeanization of domestic public policies, this book analyzes domestic systemic-level change as a dependent variable and empirically tests it by analyzing the shifts in five indicators of Europeanization: policy structure (bureaucratic (re-)organization), policy methods, policy content (normative adaptation), policy content (legislative adaptation), and policy implementation. This approach will help the researcher to empirically detect and validate domestic systemic change by using analytical indicators.

The first indicator, that is, change in policy structure (bureaucratic (re-) organization), analyzes whether there is establishment of new institutions or significant alterations to the existing ones in the policy area under analysis. Establishment of new institutions signifies transformation and domestic change, whereas administrative deadlock signifies inertia. Policy-making within already existing institutional structure signifies complete fit between domestic institutional system and the requirements of the EU, and that no change is needed.

The second indicator, that is, change in policy methods, analyzes whether there is establishment of new methods of governance or significant alterations to the existing ones in the policy area under analysis. While establishment of new methods of governance signifies transformation, continuation of the old and undefined methods of policy-making signifies inertia.

The third indicator, that is, normative adaptation, analyzes whether there is discernible adaptation of domestic policy norms to the EU policy norms. Domestic change or transformation in this case is signified by the establishment of new policy norms and strategies in the policy area under analysis. A resistance to new policy norms would signify inertia. In the case of complete fit between domestic and EU norms and strategies, there would be no adaptational pressure, thus no change is needed.

The fourth indicator, that is, legislative adaptation, analyzes whether there is discernible adaptation of the domestic legislation to the EU legislation. Here we should note that European directives have a direct impact on the content of member state policies and often lead to the transformation of national regulation. The EU-membership requires complete adoption of the *acquis communautaire*, so in this case, not much diversion or inertia is expected. If there is complete fit between the national and EU legislation, a fence-sitting attitude would be expected, not much change required. In the case of misfit, establishment of new legislation and the amendment of the existing legislation would be the expected result, and this would signify a pace-setting attitude resulting in domestic change.

The final indicator is the change in policy implementation. This study underlines the significance of the implementation of the policy objectives and the policy legislation that is acquired as a result of policy transfer. I argue that the extent of the operationalization of policy objectives, strategies, and legislation constitutes the final indicator of the domestic change. Based on this understanding, an analysis of the implementation phase provides the validation of the completeness of policy transfer. Through this five-step analysis, the reader may grasp how environmental norms and ideas turn into legal rules and institutional practices, and then how they are implemented in the real-life circumstances.

METHODOLOGY

The aim of this book is to analyze how domestic public policy interacts with the European-level policy. When studying this question, other research questions are also relevant and supportive: How can we empirically validate the impacts of Europeanization? What are the empirical indicators of domestic structural change? Can we propose a new analytical framework to study domestic change? How can we apply this analytical framework on an EU member state? What would be the empirical and theoretical conclusions?

This book takes the Czech Republic as its case-study area, and attempts to answer these questions by focusing on two significant areas of environmental policy: water management and waste management policies. These are the key policy areas which are directly linked to the environmental pollution and the health of citizens. These policy areas are significant in the discussion and evaluation of the environmental policy, because both the threats and outcomes are apparent in these policy areas.

This book focuses on the environmental policy for a variety of reasons. First of all, this policy area has been remarkably developed at both the supranational level and the national level in the recent decades. We have witnessed an impressive widening and deepening of European policy competences in the environmental policy. There are several environmental action programs in which the EU formulated its ambitions and principles for the development of environmental policy. This development of a formal set of policy principles, norms, and procedures for European environmental policy has been accompanied by a significant growth in European legislation. Certainly, this bulk of supranational legislation impacted on the member states and resulted in considerable change of national policies.

Another justification of the selection of this policy area is that European environmental policy reflects a high degree of regulatory approach, which can be detected in different regulatory strategies and policy instruments. These strategies and policy instruments have varying implications on domestic institutions and policies, and therefore create detectable transformation patterns in the member states.

Among other EU member states, the Czech Republic constitutes an interesting research avenue for a study of the development of environmental policy because it lived through drastic changes in politics, economics, and society in the post-communist period which fundamentally influenced and changed its institutional and policy frameworks in environment. When people think about the Czech Republic, they often remember air and water pollution, uncontrolled waste disposal, and devastation of the countryside. These negative environmental conditions were associated with human-health effects and serious forest damage. In terms of pollution, the former Czechoslovakia was perhaps the most adversely affected country in Europe. The Black Triangle, the heavily industrialized area where Poland, Germany, and the Czech Republic meet, was packed with chemical plants, refineries, steam-heating and power plants, and lignite strip mines. At the beginning of the 1990s, water pollution was another pressing issue of the Czech environment. The major waterways were polluted, and groundwater was highly contaminated. Regarding the waste management, hundreds of landfills were in operation until 1996, when they were forbidden by law. In addition, more than 5,000 municipalities did not have sewer systems and wastewater treatment facilities, and soil and groundwater at thousands of localities were contaminated, as a consequence of previous industrial, military, and agricultural activities.[33]

A significant improvement started with the Velvet Revolution in November 1989 that began the transition from the communist past. In this phase, there was an open window of learning for all these states in almost all public policy areas; and the political elite and the policy-makers in these states tend to search for possible solutions to the immediate problems. In such knowledge-insecurity situation, where political elites no longer knew which instruments they should use, the tendency was to imitate successful examples which can help them to emerge from their collectively felt cognitive dissonance. An additional, important long-term motivation for making positive environmental changes occurred in 1993, when the Czech Republic became a candidate country for EU accession. The substantial improvement of environmental conditions in the Czech Republic started when the Czech political elites and policy-makers had to initiate significant domestic change and adaptation to the EU through the implementation of the *acquis communautaire*. In May 2004, the country officially became an EU member state. The EU impact on the Czech Republic's environmental policy has been both direct and indirect. The direct effects have been the transfer of knowledge and financial support from the EU. These were powerful driving forces for environmental improvement in the region during the pre-accession period.[34] However, informal effects were of even greater importance; because from the beginning, politicians, state officials, and business leaders were aware that EU accession was near and had planned their actions accordingly. As a consequence, "the former communist country witnessed remarkable improvements in environmental quality in the 1990s, and they had a lasting effect on the county."[35]

Therefore, an analysis of the domestic institutional change in the Czech Republic as a result of Europeanization provides a good amount of empirical data in temporal and comparative frameworks regarding the outcomes and outputs of this policy adaptation.

The final issue to be discussed in this part is the classification and the use of empirical data. The empirical research in this book focuses on the impact of Europeanization on water and waste management policies of the Czech Republic, and this question is to be analyzed by using secondary data resources.

Domestic change in the institutional structure is empirically measured and validated through the analysis of central, regional, and municipal administrative structures in the Czech Republic. The institutional structure in environment is analyzed by using various ministerial web-pages, ministerial information brochures, and reports with an aim to find out the establishment of new institutions and new methods of governance, or the significant modification of the existing ones. Domestic normative change is analyzed through content analysis of the executive portfolio, that is, Strategy Papers published by the relevant ministries of the Czech Republic and by the European Union. Domestic legislative change is analyzed through a comparative content analysis of the EU legislation and the Czech national legal documents in water and waste management areas that appeared after the commencement of the EU legislation. In this context, existing laws, newly introduced laws, and the amendments are thoroughly analyzed with an aim to demonstrate the domestic change in the legal framework as a result of the impact of EU rules and regulations. As a consequence of these instances of domestic change, operationalization of the new policy norms and legislation is analyzed by using national and European statistics. In this context, only the official statistics that are provided by the relevant governmental agencies of the Czech Republic, by the European Union, and by verifiable international sources are used to analyze and compare the changes in these policy areas. The time period under analysis is from 2004 to 2018—when possible to acquire statistical data, the time period of analysis extends to 1991, to the post-communist era. Statistical changes in the aforementioned indicators in the water and waste management policy areas within this time period demonstrate, and thus empirically validate the operationalization and fulfillment of the policy transfer.

RELEVANCE OF THIS RESEARCH: IS IT *REALLY* IMPORTANT?

As discussed earlier in this chapter, most theories of Europeanization studies lack causal mechanisms of national adaptation. In the existing literature, attempts of empirical assessment of development of policy norms, strategies, and processes are rather scarce. In fact, comparison of policy strategies, degree of policy implementation, and comparison of policy achievements seem almost beyond feasibility. One of the aims of this book is to fill a major methodological caveat in the studies of Europeanization. This research puts Europeanization in context by establishing a consistent analytical framework for analyzing the policy processes at work.

Another relevance of this research is that it attempts to introduce a comparative outlook of environmental policy norms, rules, and strategies in the European Union and in the Czech Republic. By doing so, it demonstrates the development of environmental awareness in water and waste management policies and analyzes the reflection of this awareness into the legislative and institutional structures in both contexts. The reader may grasp how environmental ideas turn into legal rules and institutional practices, and then how they are implemented in the real-life circumstances. The end point of this long process would be seeing and understanding the benefits to environment and human health through implementation of these environmental norms, rules, and procedures. This story is told only in two environmental policy areas and in one national context, but the interested reader might search for more examples.

There is also a need for word of caution here. We must take into consideration that domestic responses to Europeanization are not static phenomena. Europeanization processes are constantly in motion and so are the domestic adaptations to them. The analytical framework proposed here does not suggest a static picture of Europeanization and domestic change. Rather, it is suggested as a tool to enable systematic empirical research on the domestic impact of Europeanization.

In further researches, same analytical framework may be applied to other EU member states and/or to other policy areas. Such studies in different contexts may provide a comparative insight and serve as a confirmation of the applicability of this analytical framework.

OUTLINE OF THE BOOK

Chapter 1 sets out the analytical framework for the whole study, reviewing the literature on Europeanization, and introduces the main explanatory concepts which will be used in the case-study policy areas.

Chapter 2 demonstrates the administrative framework in environment in the Czech Republic. It tries to find traces of institutional isomorphism as a result of the Europeanization of public policies, and tries to answer the crucial questions about the adoption of new policy methods. In this context, it identifies both the old institutions and the newly emerging institutional forms of coordinating environmental policy, and analyses structural change in this context. Then, the impact of the new administrative procedures introduced by the EU in environmental policy upon those of the Czech Republic is discussed to find out the traces of change in domestic policy methods.

Chapters 3 and 4 present case studies of two sub-fields of environmental policy: water management and waste management. These chapters provide a vast analysis of norms, strategies, legislation, and outcomes in these policy areas. First of all, I identify domestic and European environmental policy norms, and then, I comparatively analyze how they are framed within policy strategies. Secondly, I analyze legislative developments in the European Union and in the Czech Republic in water and waste management policies in a comparative perspective and in a chronological order to demonstrate domestic legislative adaptation to the EU. This part of research allows us to demonstrate varying reactions and modes of adapting to the legal constraints imposed by the European Union. Finally, I provide an empirical analysis of the policy implementation—that is, an analysis of the actual operationalization and fulfillment of policy norms, strategies, and legislation—based on statistical data. Empirical testing of the periodic change in policy outputs provides the validation of the completeness of the policy transfer.

Chapter 5 presents the conclusions and explains how the EU influence worked in the Czech Republic in the water management and waste management policy areas.

NOTES

1. Knill, C. and Lehmkuhl, D. (1999) "How Europe Matters: Different Mechanisms of Europeanization", *European Integration Online Papers* 3(7).
2. Olsen, Johan P. (1996) "Europeanization and Nation-State Dynamics," in Gustavsson S. and Lewin L. (eds.), *The Future of the Nation-State*, London: Routledge; Börzel, T. and Risse, T. (2003) "Conceptualizing the Domestic Impact of Europe," in Keith Featherstone, Claudio Radaelli (eds.), *The Politics of Europeanisation*, Oxford: Oxford University Press; Knill and Lehmkuhl (1999) "How Europe Matters"; Tanil, Gamze (2012) *Europeanization, Integration and Identity*, New York: Routledge.
3. Héritier, A., Kerwer, D., Knill, C., Lehmkuhl, D., Teutsch, M. and Douillet, A.C. (2001) *Differential Europe*, Lanham: Rowman and Littlefield, p. 3.
4. Radaelli, C. (2003) "The Europeanization of Public Policy," in Keith Featherstone, Claudio Radaelli (eds.), *The Politics of Europeanisation*, Oxford: Oxford University Press, p. 30.
5. Ladrech, R. (1994) "Europeanization of Domestic Politics and Institutions: The Case of France," *Journal of Common Market Studies*, 32(1), p. 69.
6. Ibid, p. 71.
7. Ibid, p. 72.
8. Olsen (1996) "Europeanization and Nation-State Dynamics"; Börzel and Risse(2003) "Conceptualizing the Domestic Impact of Europe."
9. Risse, T., James, C. and Cowles, M. (2001) "Europeanization and Domestic Change: Introduction", in Maria Green Cowles, James Caporaso and Thomas Risse (eds), *Transforming Europe: Europeanization and Domestic Change*, Ithaca, NY: Cornell University Press, p. 1.
10. Héritier et al. (2001) *Differential Europe*, Lanham: Rowman and Littlefield, p. 3.
11. Börzel and Risse (2003) "Conceptualizing the Domestic Impact of Europe"; Risse et al. (2001) "Europeanization and Domestic Change"; Börzel (1999) "Towards Convergence in Europe?"; Knill and Lehmkuhl (1999) "How Europe Matters"; Knill and Lenschow (2001) "Adjusting to EU Environmental Policy: Change and Persistance of Domestic Administrations"; Tanil (2012) *Europeanization, Integration and Identity*.
12. Börzel and Risse (2003) "Conceptualizing the Domestic Impact of Europe," p. 61.
13. Risse et al. (2001) "Europeanization and Domestic Change," p. 7.
14. Knill and Lehmkuhl (1999) "How Europe Matters," p. 9.
15. Ibid., p. 16.

16. Dolowitz and Marsh (1996), "Who Learns What From Whom: A Review of the Policy Transfer Literature," *Political Studies* XLIV, p. 343–357

17. Radaelli, C. (2000) "Policy Transfer in the European Union: Institutional Isomorphism as a Source of Legitimacy," *Governance* 13(1), p. 25–43.

18. Dolowitz and Marsh (1996), "Who Learns What From Whom: A Review of the Policy Transfer Literature," p. 350.

19. Bennett (1991) "What is Policy Convergence and What Causes It?", *British Journal of Political Science* (21)2, p. 218.

20. Ibid., p. 218.

21. Knill and Lehmkuhl (1999) "How Europe Matters," p. 4.

22. Ibid., p. 4.

23. Ibid., p. 4.

24. Ibid., p. 4.

25. DiMaggio, P.J. and Walter W.P. (1991) "The Iron Cage Revisited: Institutional Isomorphism and Collective Rationality in Organizational Fields", in Walter W. Powell and Paul J. DiMaggio (eds.), *The New Institutionalism in Organizational Analysis*, Chicago, London: University of Chicago Press, p. 67.

26. Ibid., p. 67.

27. Ibid., p. 69.

28. Bennett (1991) "What Is Policy Convergence and What Causes It?" p. 221.

29. Ibid, p. 221.

30. Ibid, p. 224.

31. Ibid, p. 225.

32. Ibid., p. 227.

33. Ministry of Environment (1996) "Report on the Environment in the Czech Republic," Prague.

34. Reiche, D. (2006) "Renewable Energies in the EU-Accession States," *Energy Policy*, 34, pp. 365–375.

35. Moldan, Bedrich (2007) "Environment in the Czech Republic: A Positive and Rapid Change," *Environmental Science and Technology*, (41)2, p. 359–362.

Chapter 2

Institutional Structure in Environment

This chapter presents the domestic institutional structure of the Czech Republic in the environmental policy area. In this chapter, I look at the existing and newly emerging institutional forms of coordinating environmental policy such as environmental state agencies, national, regional, and municipal environmental organizations, and the newly emerging policy procedures when dealing with the environmental policy.

EU member states establish new institutions or alter the existing institutions significantly in order to respond to the European-level institutional and policy developments and to enable the transfer of policy methods. As an example to such analyses, Featherstone[1] describes Europeanization based on how public administrative institutions adapt to the obligations of EU membership. Considering that the European-level institutional and policy developments bring states and the EU into closer contact with one another, Rometsch and Wessels[2] refer to this process as "institutional fusion." They put forward that "national institutions have been deeply Europeanized, meaning that they have adapted their structures and procedures and are, to a growing extent, oriented towards the activities in Brussels and Strasbourg."[3]

However, it should be also noted that the institutions change only slowly, and this change might be far less obvious than the changes in national policies. Confirming this view, Bulmer and Burch stated that "at the level of machinery, governmental structures and procedure, the impact of Europe has been far less evident than on the substance of policies."[4]

Notwithstanding the fact that the institutional change emerges slowly and less obviously than the changes in policies, political scientists still try to explain its mechanisms. Institutional isomorphism is one of the mechanisms of institutional change. If we consider that "organizations compete not just for resources and customers, but for political power and institutional legitimacy,"[5] then the motivation for copying other institutional structure can be found in the efficiency considerations as well as in the search for legitimacy in political life.[6]

Especially when there is dissatisfaction with the status quo or an obvious problem to be overcome, then the domestic policy-makers may resort to institutional isomorphism much easier: "When government policies are functioning properly there is no need to search for lessons; everything can operate through established routines. Only when routines stop providing solutions, is it necessary to search for lessons."[7] Therefore, it might be the situations of uncertainty that encourage imitation. In such situations, "mimetism stems from the need to cope with uncertainty by imitating organizations perceived to be more legitimate or more successful."[8] When the environment creates symbolic uncertainty, mimetism provides an easy and cheap way of finding quick solutions to immediate problems. In this process, the modeled organization may be unaware of the modeling or may have no desire to be copied; rather, it merely serves as a convenient source of practices that the borrowing organization may use.

This mechanism can be particularly apparent in the post-communist context, where policy-makers were under great pressure to find models quickly in order to replace discredited communist institutions. Indeed, Knill and Lenschow argued that "core (institutional) changes may occur as the result of external shocks, such as the political and economic transition we witnessed in the CEE states."[9] In the situations of crisis, there is a natural tendency to look abroad, to see how other states have responded to similar pressures, to share ideas, to draw lessons, and to bring foreign evidence to bear within domestic policy-making processes. When new policy problems require innovation, "emulation can have a salutary effect in an insecure and tentative policy-making climate";[10] and when more urgency is perceived, "the more likely will be the imitation of solutions without lengthy analysis and investigation."[11]

The resonance between the structures eases the institutional isomorphism: "policy-makers might be willing to emulate some countries more than others, especially ones that display more institutional compatibilities."[12]

The motive for institutional isomorphism does not only reside in the needs of the domestic policy-makers. Especially after the EU accession of new member states, European institutions may also stimulate changes in the domestic institutional structure. Given that powerful organizations may exert both formal and informal pressures upon their dependent organizations, domestic structures might be subject to the influence of a powerful EU. DiMaggio and Powell conceptualize this mechanism as a *coercive process*[13] because it stems from political influence and as the dependency increases, mimetism in structure and behavior increases as well.

On the opposite side of the coercive pressures, we can find *normative pressures* that stem from professionalization. What happens at the European level can be described as a production of a common cognitive base and a shared legitimization of an occupational autonomy by the professionals and their associations through the mechanisms of socialization.[14] These normative pressures may encourage organizational structures to resort to institutional isomorphism.

Based on this theoretical background, this chapter tries to find traces of institutional isomorphism in the Czech Republic's institutional structure in the environment policy as a crucial impact of its accession to the EU.

A BUREAUCRATIC (RE-)ORGANIZATION?

Analysis of the Institutional Structure in Environment

At the central level, Ministry of the Environment (MoE) is the central authority for the environmental policy. Established on January 1, 1990, the Ministry administers the protection of natural water accumulation; protection of water resources and the quality of groundwater and surface water; air protection; nature and landscape protection; conservation of agricultural land; operation of the National Geological Survey; protection of the mineral resources and groundwater; environmental supervision of mining, waste management, fisheries, and forestry in national parks; environmental impact assessment of activities and their consequences; and the making of national environmental policy.[15] The organizations within the jurisdiction of the Ministry are Cave Administration, Czech Environmental Information Agency, Czech Environmental Inspectorate, Czech Geological Survey, Czech Hydrometeorological Institute, and State Environment Fund of Czech Republic.

Agency for Nature Conservation and Landscape Protection of the Czech Republic (AOPK CR) was established by the MoE on March 1, 1995[16] as a successor of the former Czech Institute for Nature Conservation. The main aim of AOPK is to protect and conserve nature and landscape on the territory of the Czech Republic. The main responsibilities of the Agency are:

> monitoring the status, changes and trends in selected habitats and populations of the endangered and specially protected wildlife species; technical and expert support to the state administration; administration of the Nature Conservancy Central Register; implementation of practical measures to conserve nature and protect landscape in the 25 Protected Landscape Areas and 216 National Nature Reserves and National Nature Monuments on the whole Czech Republic's territory; awareness raising and education; and international cooperation.[17]

Czech Environmental Agency (CENIA) was established on April 1, 2005,[18] under the framework of the MoE with a mission to collect, evaluate, interpret, and distribute the environmental information. The role of this agency is to ensure the fulfillment of the mandatory reporting of environmental data and to provide crosscutting environmental information.[19] CENIA cooperates with all providers of data sources in the environmental sector as well as with a number of research institutes, scientific centers, or universities. It participates in the development and provision of selected data and map services and operates a number of information systems. CENIA uses its expertise in the preparation of national evaluation documents and strategic and conceptual materials produced by the MoE.

In the Czech Republic, CENIA is the contact point of the European Environment Agency (EEA) and is involved in the Eionet European Environment Information and Observation Network. It is also the national contact point for the Infrastructure for Spatial Information in Europe INSPIRE and the representative body of the European Ecolabel program—Ecolabel EU and the eco-management and audit scheme (EMAS).[20]

The Czech Environmental Inspectorate (CEI) was set up in 1991[21] as an expert executive body within the state administration to supervise the environmental legislation enforcement. It is responsible for ensuring compliance with environmental law, acting under the aegis of the MoE. It consists of a central directorate in Prague and ten regional branches. The activities of CEI can be divided into five core areas of environmental policy: air protection, water protection, waste management, nature protection, and forests protection. It conducts inspections and controls; prescribes measures to remedy identified shortcomings; imposes fines for non-compliance with environmental legislation; and imposes restrictions of operations that pose a serious danger to the environment.[22] Regarding the international cooperation, CEI works in close cooperation with the EU Network for the Implementation and Enforcement of Environmental Law (IMPEL),[23] and takes an active part in working of its expert groups and projects.

The State Environmental Fund of the Czech Republic (SEF CR), founded in 1992,[24] substantively contributes to investments aimed at protecting and improving the environment. The Fund co-finances projects to improve the quality of water, air, and waste management. It provides financial support through subsidies, soft loans or a combination of both. The Fund obtains the financial resources from the state budget and from polluters. Polluters pay fees for discharges of wastewater, for land reclassification, air pollution fees, and fees levied under the Waste Act. The Fund is also responsible for administering funds from the EU Structural Funds and the Cohesion Fund in the environmental sector in the Czech Republic.[25]

Former Minister of Environment, Bedrich Moldan perceives the Fund as an important actor in environment considering that "it administers both the state budget on environment and EU budget on environment."[26] As an illustration, in 2015, total expenditure of the Fund amounted to CZK 3,111 million. Expenditure related to the implementation of projects focused on water protection had the largest share (33.4%) with CZK 1,040 million. Expenditure related to the implementation of projects focused on waste had a smaller share (9.5%) with CZK 2,394 million.[27]

T.G. Masaryk Water Research Institute (TGM WRI) was founded as the State Hydrological Institute in 1919.[28] In the 1950s, the hydrological and meteorological services were separated from the research and they were subsequently moved to a newly founded Hydrometeorological Institute. Research on surface water, groundwater, and precipitation hydrology remained in the Institute. After 1989, the Institute was transferred to the supervision of the newly established MoE. In 1999, the Institute's activities were extended to waste management. Since January 1, 2007, the status of the Institute is a public research institution.[29]

The activities of TGM WRI include research on the status, use, and changes of aquatic ecosystems and their linkages in landscape and related environmental risks, waste, and packaging waste management, as well as professional support to water protection, flood prevention, and management of waste and packaging waste.[30]

The Czech Geological Survey (CGS) was established in 1919 with the task of providing geological service for the Czech Republic.[31] Although the character of the institution, together with its name, has changed over time, its main mission and related unique social status have remained. The CGS has the statutory responsibility to gather, store, and interpret geological information so that the state administration can take appropriate decisions about national economic and environmental issues. It provides the results of systematic regional geological mapping and investigation to all interested persons.

The Czechoslovak Meteorological Institute was established in 1919. In 1954 this meteorological institute was combined with hydrology, establishing today's hydrometeorological institute.[32] The amended law in 2015[33] established the Czech Hydrometeorological Institute (CHMI) as "a semi-autonomous state organisation serving in the fields of air quality, hydrology, water quality, climatology and meteorology, and providing services for the state administration, private sector and general public" (Article 2).

Considering the recent concern about climate change, the CHMI offers a mechanism called *Clidata*, which is of Czech origin. In the area of climatology, it is important to learn patterns and phenomena in the climate system in order to provide a detailed analysis of the climate data for the measurement period. The behavior of the climate system cannot be described unless reliable and accessible groups of climate elements have been available for a long time. To achieve this goal, the CHMI developed the Clidata system intended for archiving of climatology data. The System was designed to replace the old Clicom system, which had been used in Czech Republic from 1993 to 2000. The new Clidata system has been used by the CHMI ever since and it is successfully installed in more than thirty other countries.[34]

Under the aegis of the Ministry of Environment, the Government Council for Sustainable Development was established in 2003[35] as an advisory, initiative, and coordinating body in sustainable development and strategic management. The Council is chaired by the prime minister, and is responsible for inter-sectoral coordination of the sustainable development policy among central administrative authorities. Its nine thematic committees[36] and working groups bring together representatives of all ministries, parliament, municipalities, nongovernment organizations, trade unions, industry, and academia.[37] Main responsibilities of the Council are the development of the sustainable development strategy and its reviews; biannual reporting on the implementation of the sustainable development strategy in the Czech Republic and monitoring of indicators; and methodological coordination of conceptual documents. In 2014, the Government Council for Sustainable Development was moved from the MoE to the Government Office.

In the post-communist period, reform of the territorial public administration resulted in the establishment of regions[38] and in the transfer of a number of competencies from the central state administration to the regional administrations. The Czech Republic was divided into 14 regions[39] and 6,258 municipalities. The competences of the regions in the environmental policy area were designated as decision-making on categorization of forests; leading and elaboration of waste register; approving of treatment with hazardous substances; and performance in the area of protection of nature, air, and agricultural land resources. The authorities of the municipalities were designated as water management, administration of small streams, flood-management, flood prevention, waste management and waste treatment, protection of nature and air, and protection of agricultural land resources.

Conclusive Remarks

The public administration in the Czech Republic went through profound transformation in the post-communist period. Shortly after the post-communist transition, main body of the institutional structure was established regarding the environmental policy-making, policy-implementation, and policy inspection in the Czech Republic. This analysis demonstrated that the post-communist transition was more influential on the bureaucratic (re-)organization of the environmental institutional structure of the Czech Republic than the EU accession.

An important theoretical hypothesis of this chapter was that in the situations of crisis, there is a natural tendency that the political elite would look abroad to draw lessons and to bring foreign evidence to bear within domestic policy-making processes. When new policy problems require innovation, institutional isomorphism through mimetic processes provides an easy and cheap way of finding quick solutions to immediate problems.

When asked about their experience during the post-communist transition phase, the first Minister of Environment of the Czech Republic, Bedrich Moldan, argued that "in the 1990s we looked at how environmental legislation in other European countries are made, and how the EU environmental legislation was transposed into their national legislation."[40] In the communist period, "there was an environmental administration, but not a specialized ministry for environment; instead there were different bodies acting in this policy area but in an uncoordinated way, and the environmental legislation was not good. So, they had to change many things deeply."[41] When asked about a specific example to these fundamental changes, Mr. Moldan provided that "In the communist period, there were two separate institutions on water management and forestry. When the Ministry of Environment was established, we cancelled former institutions and merged their competences under the Ministry of Environment." He provided another specific example of a fundamental change in water management: "The biggest development (in water management) has been the change away from the old communist institutional structure. This change was not connected to the EU. The EU-membership only provided gradual improvement of the system."[42]

Therefore, the influence of the post-communist transition was stronger on the bureaucratic (re-)organization of the environmental institutional structure of the Czech Republic than the influence of the EU accession. When asked about his perception and definition of the impact of EU accession on the Czech Republic's institutional structure in environmental policy, Mr. Moldan stated that "institutional structure in the environmental policy was created in the early 1990s; so, the bulk of the institutional (re-)organization did not happen as a result of the EU-accession."[43]

These empirical findings demonstrate that the process of European integration had lesser impact upon the institutional structures of Czech environmental policy-making than the post-communist transition. However, it is also undeniable that the domestic institutional structure has evolved and improved simultaneously with the developments in the EU environmental policy.

NEW POLICY METHODS

Analysis of the Policy Methods in Environment

In this section, I will analyze the new administrative procedures and methods introduced by the EU in environmental policy by categorizing them under five areas:

- Assessment procedures,
- Information procedures,
- Planning procedures,
- Auditing procedures,
- Sanctioning procedures.

In the context of the assessment procedures, two methods were introduced by the EU: environmental impact assessment in 1985 and strategic environmental assessment in 2001. These new policy methods obviously had policy implications on the member states. In the context of the methods for public information, the EU adopted the Directive on Free Access to Environmental Information in 1990, and INSPIRE Directive in 2007. These had obvious policy implications in the member states. As for the planning procedures, the EU introduced four methods specifically in the water management and waste management areas: river basin management plan; water supply and sewerage systems plan; waste management plan; and waste prevention program. The implementation of these methods in the member states created different degrees of adaptational pressures. Regarding the auditing methods, the EU developed a voluntary Eco-Management and Auditing System in 1993; so, its impact on the private companies in the Czech Republic is analyzed here. Finally, the EU introduced criminal sanctions for environmental violations in 2008, which became a new step in the area. This move created heavy influence on the member states by leading them to change their domestic system of sanctioning in the area of environment, and adopt a new mentality in the case of environmental hazards.

Assessment Procedures

In 1985, the European Union adopted the Environmental Impact Assessment Directive (EIA Directive),[44] which applies to the assessment of the environmental effects of those public and private projects which are likely to have significant effects on the environment (Article 1). EIA Directive obliges developers of specified kinds of public and private projects to provide information regarding the environmental impact of these projects to a designated public authority in the area of the environment (Article 1). This environmental impact assessment must then be taken into consideration by those public authorities that are responsible for the authorization of the project(s) in question.[45]

Considering institutional arrangements, the EIA Directive imposes a procedural requirement on public and private investors as well as on regulatory authorities, but it does not impose any standards regarding the conditions under which a project should be denied authorization. The Directive also implies an increase of control responsibilities of the administrative structures. Since an impact assessment deals with the direct and indirect effects of the project on human beings, fauna, flora, soil, water, air, climate, landscape, material assets, and cultural heritage (Article 3), the designated public authority ought to be able to evaluate these impacts in a comprehensive way, hence signifies a need for integrating administrative structures.

The impact of the Directive on domestic administrative procedures of the member states is unavoidable. Article 2.2 states that "the EIA shall be integrated into the existing procedures for consent to projects in the Member States, or, failing this, into other procedures or into procedures to be established to comply with the aims of this Directive." Member states are obliged to take the measures necessary to comply with this Directive within three years of its notification (Article 12). Thus, assuming that a member state did not have a similar procedure before (existence of an institutional misfit), transposition of this Directive would be expected to trigger a change in domestic institutional methods.

The Environmental Impact Assessment process was transposed into the Czech Republic's domestic structure in 1992.[46] The process constituted both an important element of preventive environmental protection instruments and a significant component of the environmental policy. In 2001, a new act[47] was adopted to fulfill the requirements of the amended EU Directive. This act regulated "environmental and public health impact assessment and the procedure of natural persons, legal entities, administrative bodies and territorial self-governing units (municipalities and regions) in this assessment" (Article 1.1). The scope for the assessment covers the project's impacts on "population and public health and the environment, including animals and plants, ecosystems, biodiversity, soil, water, air, climate and landscape, natural resources, tangible assets and cultural heritage" as defined by the legislation (Article 2). Method of assessing the effects of the project on the environment shall include the "identification, description, assessment and evaluation of the anticipated direct and indirect significant effects of implementation and non-implementation of the project" (Article 5.1). When assessing a project, "the effects on the environment during its preparation, implementation, operation and its termination, eventually the consequences of its liquidation and further remediation or reclamation of the territory shall be evaluated" (Article 5.3).

The assessment of the project shall also include "a proposal for measures to prevent possible significant negative effects on the environment through the implementation of the project, as well as measures to eliminate, reduce, mitigate or minimize these impacts, or measures to increase the positive effects on the environment by carrying out the project" (Article 5.4).

However, the 2001 Act had shortcomings related to public participation in decision-making and access to justice in environmental matters. The case was transmitted to the European Court of Justice. A bill amending the act was passed in 2009, but the public participation issues were not solved, and the European Commission opened a second infringement procedure. In 2012, the Aarhus Convention Compliance Committee reiterated what it saw as the act's deficiencies.[48] Under the threat of heavy financial sanctions for non-compliance, the Czech Republic made complying with the EIA Directive a priority and approved a new bill amending the act, which took effect in April 2015. The 2015 amendment,[49] which addressed the infringements at that time, still raised issues. Therefore, September 2017 amendment[50] focused on streamlining and shortening the procedure. Activities subject to EIA were divided into two categories: projects with expected high impact, such as large energy and infrastructure projects, must undergo mandatory EIA, while those with less potential impact are only subject to EIA following a decision of the competent authority. Depending on the type of project, the competent authority is either the Ministry of Environment or a regional authority. Screening is based on information provided by the project proponent regarding the planned activity and the local state of the environment.

The Strategic Environmental Assessment Directive (SEA Directive)[51] was adopted by the EU on June 27, 2001. The objective of the Directive was to ensure that an environmental assessment is carried out for certain plans and programs which are likely to have significant effects on the environment (Article 1). The SEA Directive applies to a wide range of public plans and programs which are prepared for agriculture, forestry, fisheries, energy, industry, transport, waste management, water management, telecommunications, tourism, and land use (Article 3.2.a). Whereas EIA aims to give decision makers information on the environmental impact of all real courses of action and to propose the most suitable option of solution, SEA aims to contribute to the integration of environmental considerations into the preparation and adoption of strategic planning documents.[52] The impact of the Directive on domestic administrative procedures is unavoidable since it states that "the requirements of this Directive shall either be integrated into existing procedures in Member States for the adoption of plans and programmes or incorporated in procedures established to comply with this Directive" (Article 4.2).

The SEA Directive is transposed into the Czech institutional structure by an amendment[53] to the Act on Environmental Impact Assessment. This is a framework act for both EIA and SEA procedures. The SEA procedure is regulated by Article 10a-10j of this Act. The process of strategic environmental impact assessment is based on the systematic examination and assessment of the potential environmental impact. The purpose of this is to determine, describe, and carry out comprehensive evaluation of the expected impacts of prepared plans (strategies) on the environment and public health in all decisive contexts.

Several authorities are involved in the SEA procedure in the Czech Republic. The competent authorities are the Ministry of Environment or Regional Authority in whose territorial jurisdiction a plan or program is being drawn up. These authorities are responsible for leading the SEA procedure. The concerned administrative authority is the authority that represents the interests protected by the special legal regulations and whose jurisdiction at least partly corresponds to the affected territory. In all cases, one of the concerned administrative authorities is the CEI.

Information Procedures

On June 7, 1990, the EC adopted the Directive on Freedom of Access to Information on Environment[54] with an aim to ensure freedom of access to, and dissemination of, information on the environment held by public authorities and to set out the basic terms and conditions on which such information should be made available (Article 1). To make the activities of both public authorities and the regulated industries accountable to the public, the Directive required that relevant authorities holding information on the environment make this information available to the persons requesting it (Article 3.1).

This Directive established a passive institutional procedure, because the public and private authorities are obliged to respond to requests, not to proactively engage in provision of information. This means that the institutional implications of the Directive depend on the public demand for information. The procedural impacts of the Directive affect patterns of administrative style, since it demands open and transparent channels between the general public and the administration.

In 1998 the Czech Republic adopted the Law on Access to Information on the Environment,[55] which stipulated the conditions of exercising of the right to timely and complete information on the state of the environment and natural resources (Article 1). Article 3 stated that "For the purpose of obtaining information about the environment the applicant can ask the body to make the information about the environment accessible. He need not give grounds for his application. The request can be made in the oral, written, telephone, telegraph, fax or other technically feasible form." Upon this application, the information shall be made available as soon as possible, at the latest within thirty days of the request, unless special circumstances justify an extension of this period, in the extreme case up to sixty days (Article 7).

In addition to the Law on Access to Information on the Environment, the Czech Republic adopted the Law on Free Access to Information[56] in 1999 which obliged state authorities and communal bodies to provide information related to their activities (Article 2). Article 4 of the Directive stated that the obliged entities provide information to the applicant on the basis of their application or by disclosure.

In the EU, a new Council Directive on Public Access to Environmental Information[57] repealed the former Directive on Freedom of Access to Information on Environment. Article 7.1 of the new Directive stated that member states shall ensure that environmental information progressively becomes available in electronic databases which are easily accessible to the public through public telecommunication networks. In addition, Article 7.2 specified which information shall be made available and disseminated. These shall include policies, plans, and programs relating to the environment; reports on the state of the environment; and data or summaries of data derived from the monitoring of activities affecting, or likely to affect, the environment.

To provide compatibility with the Article 7.1 of this Directive, the Report on the Environment of the Czech Republic is drawn up every year on the basis of Act No. 123/1998 Coll., on the right to information on the environment. The Ministry of Environment of the Czech Republic publishes the annual environmental reports and cooperates with regional authorities to issue a similar report for each region, accompanied by a comparative analysis of environmental indicators across regions. These reports are published on the Ministry of Environment website. It is a comprehensive document assessing the state of the environment in the Czech Republic, including its entire context, on the basis of data available in the given assessed year.

Starting with 2005, the Czech Environmental Information Agency became responsible for drawing up annual reports. CENIA's main publications are open to public access and they include the report on the Environment of the Czech Republic, reports on the environment in the regions of the Czech Republic (the so-called regional reports) and the Statistical Yearbook of the Environment of the Czech Republic. CENIA also provides data for the Information System of Environmental Statistics and Reporting (ISSaR), which contains a well-arranged set of all statistical data and indicators of the state of the environment. Furthermore, the Statistical Environmental Yearbook is prepared yearly by the Czech Statistical Office and published on the Ministry of Environment and Statistical Office websites.

The CGS also provides the development, maintenance, and continuous updating of the Information System of the State Geological Survey[58] providing effective access to data and information. The CGS provides mining maps database; geophysical database; underground mines database; mine land reclamation and remediation database; inventory of closed and abandoned mining waste facilities; inventory of reserves of reserved and non-reserved mineral deposits; inventory of newly commenced geological projects; and territorial data for land-use analysis.[59]

In addition to those annual reports and regularly maintained databases, details on all permits granted can also be found at the MoE website. A summarized version of every inspection report is also published on a publicly accessible information system run by the MoE. The public can also request inspection results for any regulated facility. Publicly accessible information on regulated entities' compliance records improves compliance with environmental law. The CEI website has a lot of information available to the public, in line with EU requirements for openness and transparency.

In 2007, the EU adopted the INSPIRE Directive[60] with an aim to lay down general rules aimed at the establishment of the Infrastructure for Spatial Information in the European Community for the purposes of Community environmental policies and policies or activities which may have an impact on the environment (Article 1). The Directive required member states to ensure that metadata are created for the spatial data sets and services corresponding to the themes listed in the Directive, and keep them up to date (Article 5).

The INSPIRE Directive was transposed into Czech Republic by an amendment[61] to Act No. 123/1998 Coll. on the right to information on the environment. As a result, the MoE of the Czech Republic established the INSPIRE National Geoportal. CENIA manages the INSPIRE National Geoportal, where besides the data and services covered by the INSPIRE Directive, thematic data from several dozens of entities ranging from central public authorities, departmental organizations, regional authorities, municipalities, research institutions to private companies can be found. These data sets are open to the general public, and they allow users to search, view, and download data. All data and services made available on the geoportal have metadata completed in accordance with INSPIRE requirements.

Institutional structure has also changed to allow for better coordination for spatial data. In November 2010 an INSPIRE national coordinating committee (INCOCO) was put in place with twenty-one organizations involved. It is composed of representatives of central and local public administration bodies and interest organizations associating entities operating in the field of spatial data. The aim of the committee is the coordination of data providers, strategies, and obligations toward the EU. The INCOCO secretariat is CENIA and its secretary is the CENIA director.

Planning Procedures

In 2000, the EU adopted the Water Framework Directive[62] which required each member state to establish for each river basin district, or for the part of an international river basin district within its territory, a program of measures (Article 11), which include the basic measures (Article 11.3), and where necessary, supplementary measures (Article 11.4). The programs of measures shall be established at the latest nine years after the date of entry into force of this Directive, and all the measures shall be made operational at the latest twelve years after that date (Article 11.7).

These planning methods were adopted by the Czech Republic through the Water Act[63] in 2001. The Water Act stipulated that Plan of Main River Basins of the Czech Republic to be prepared as a strategic water planning document (Section 24.1) by the Ministry of Agriculture in cooperation with the Ministry of Environment (Section 24.2). In accordance with the Council Directive, the measures for achieving the aims of water protection must include the principle measures and, where necessary, supplementary measures (Section 26.2) and they must be implemented within three years from approval of the plan (Section 26.1). The implementation of this policy method followed soon after. The Plan of Main River Basins of the Czech Republic was prepared by the Ministry of Agriculture in cooperation with the Ministry of Environment, and approved on May 23, 2007.[64]

The EU adopted Urban Waste Water Directive[65] on May 21, 1991, with an aim to regulate the collection, treatment, and discharge of urban waste water and the treatment and discharge of waste water from industrial sectors. The Directive required member states to establish a program for the implementation of this Directive by December 31, 1993 (Article 17.1), and provide the Commission by June 30 every two years with an update of the information on the program (Article 17.3).

The Czech Republic adopted the policy method of preparing national and regional development plans for water supply and sewerage systems with the Act on Water Supply and Sewerage Systems.[66] The Act stated that the Ministry of Agriculture shall ensure the development and updating of the national water supply and sewerage development plan (Article 29(1)b). Pursuant to this article, the Ministry developed the water supply and sewerage development plan. This plan included standpoints of the Ministry of Agriculture and the updates of the Regional Development Plans for Water Supply and Sewerage Systems.

The EU adopted the Waste Directive[67] for the first time in 1975. This Directive required the competent authority or authorities of the member states to draw up as soon as possible one or several waste management plans covering the type and quantity of waste to be disposed of, general technical requirements, suitable disposal sites, and special arrangements for particular wastes (Article 6). This Directive was amended several times. The recent amendment[68] in 2008 obliged member states to prepare both waste management plan (Article 28) and waste prevention program (Article 29). Both the WMPs and the WPPs should be submitted to the Commission (Article 33), and evaluated at least every sixth year (Article 30).

In 2001 the Czech Republic adopted the Waste Act.[69] First, the Act stated that the waste management plan shall be prepared for the purpose of creating conditions for the prevention and management of waste (Article 41.2). Secondly, the Act clarified the content of the waste prevention program as the waste prevention objectives in accordance with the Waste Management Plan of the Czech Republic; waste prevention measures; and a set of indicators to evaluate the achievement of waste prevention objectives and measures, which will enable monitoring and assessment of the progress made through them (Article 41(a)2).

Preparation of the Waste Management Plan of the Czech Republic was assigned to the MoE (Article 42). Regional administrations were obliged to prepare waste management plan of the region in accordance with the binding part of the Waste Management Plan of the Czech Republic (Article 43). Municipalities were also given an obligation to prepare municipal waste management plans for the waste they produce and the waste they manage on the condition that if they produce more than 10 tons of hazardous waste or more than 1,000 tons of other waste annually (Article 44.1).

In compliance with these legal obligations, the Czech Republic adopted the waste management plan[70] on July 1, 2003, to establish a sustainable waste management policy. The Plan was drawn up for a period of ten years, that is, from 2003 to 2012. The waste prevention program of the Czech Republic[71] was also prepared by the MoE, and on December 12, 2013, the Draft Program was published on the Ministry‘s website. The program was approved by the government on October 27, 2014.[72]

Auditing Procedures

The European Union created the Community EMAS[73] as a voluntary activity of enterprises intended to improve the environmental performance of the enterprise and to provide the relevant information to the public and other stakeholders. The objective of EMAS is “to promote continuous improvements in the environmental performance of organisations by the establishment and implementation of environmental management systems by organisations; the systematic, objective and periodic evaluation of the performance of such systems; and the provision of information on environmental performance” (Article 1.2). With this aim, the regulation required enterprises to prepare an environmental statement that is written in a concise, comprehensible form understandable by the public (Article 5.2). The environmental statement shall include a description of the company's activities at the site considered; an assessment of all the significant environmental issues of relevance to the activities concerned; a summary of the figures on pollutant emissions, waste generation, consumption of raw material, energy and water, noise and other significant environmental aspects, as appropriate; other factors regarding environmental performance; a presentation of the company's environmental policy, program and management system implemented at the site considered (Article 5.3). The regulation required each member state to designate the competent body responsible for carrying out these tasks (Article 18). This regulation was primarily intended for the organizations from production (industrial) sector.

The government of the Czech Republic supported the introduction of EMAS program in accordance with the EU. Thus, the EMAS regulation was transposed to the Czech Republic in 1998.[74] On the basis of Government Resolution, the Council of the EMAS Programme and the EMAS Agency were established as the responsible bodies for the EMAS Programme in the Czech Republic; and an accreditation body was established for the EMAS Programme. The Resolution also issued rules for implementation of the EMAS; the national program of implementation of the EMAS; and methodical guidelines for accreditation of the environmental verifiers. By doing so, "the Czech Republic became the first of the candidate countries to introduce a fully functional EMAS system; and in five years, the register of companies with introduced EMAS contained 480 industrial enterprises and service enterprises, of which 9 organizations has registered EMAS."[75]

When the European Union updated the program and replaced it with EMAS II, the Czech Republic was very proactive in adopting the updated version of the program.[76] EMAS II focused also on companies from other sectors. In connection with this new program, the Ministry of Environment prepared the Environmental Management Accounting Implementation Guideline which came into effect on January 1, 2003. This Guideline was created on the basis of the results of the Environmental Management Accounting Working Group which was composed of the representatives of the Ministry of Environment, enterprises, universities, and the experts from the Czech Environment Institute.

When the interest in EMAS II implementation decreased, the European Commission wanted to make an effort to improve this situation by an amendment, known as EMAS III.[77] The main objectives of the revision were to increase the number of organizations involved in the Program, and to provide the recognition of EMAS as the highest standard for the implementation of environmental management systems. In this updated regulation, basic requirements for organizations that wish to register remained the same: "organizations wishing to be registered in the system shall carry out an environmental review of all environmental aspects of the organisation; develop and implement an environmental management system taking into account the best environmental management practice for the relevant sector; carry out an internal audit; and prepare an environmental statement" (Article 4.1). Upon completing these procedures, it was stated that any organization meeting these requirements may apply for registration (Article 5.1). Member states were required to take appropriate measures to provide information to the public about the objectives and principal components of EMAS, and about the contents of this regulation (Article 34).

EMAS has not found a broad application in the Czech Republic. Currently, there are twenty-five organizations in the Czech Republic holding an EMAS register. The number of validated companies is small compared to other countries of Western Europe. The reasons for this situation can be found in that compared to EMAS register, ISO 14001 is a more widely used standard. This fact encourages the companies to use ISO 14001 standard. Currently, 4,800 companies in the Czech Republic hold ISO 14001 certification.

Sanctioning Procedures

In 2008, the EU adopted the Directive[78] on *protection of environment through criminal law*. This Directive establishes measures relating to criminal law in order to protect the environment more effectively (Article 1). The Directive declares that the following conducts constitute a criminal offence when conducted unlawful or with negligence and if they cause or are likely to cause death or serious injury to any person or substantial damage to the quality of air, the quality of soil or the quality of water, or to animals or plants:

> (a) the discharge, emission or introduction of a quantity of materials or ionising radiation into air, soil or water; (b) the collection, transport, recovery or disposal of waste, and shipment of waste; (c) the operation of a plant in which a dangerous activity is carried out or in which dangerous substances or preparations are stored or used; (d) the production, processing, handling, use, holding, storage, transport, import, export or disposal of nuclear materials or other hazardous radioactive substances; (e) the killing, destruction, possession or taking of specimens of protected wild fauna or flora species, and their trading; (f) the production, importation, exportation, placing on the market or use of ozone-depleting substances. (Article 3)

The Directive requires the member states to ensure these conducts to be punished as a criminal offence (Article 4).

In the Czech Republic, criminal sanctions for environmental violations were introduced with the 2009 Criminal Code. The criminal sanctions for environmental damage or hazard were provided for natural persons in Articles 293–308. The Criminal Code provides imprisonment for up to three years for whoever intentionally damages or endangers soil, water, air, forest, or another component of the environment in a larger extent[79] (Article 293.1). In such a case, gaining extensive profit by the act increases the term of imprisonment up to eight years (Article 293.3). The Code also provides imprisonment for up to six months for the negligent conduct creating environmental damage or hazard (Article 294.1). The Code provides criminal sanction for the unauthorized waste disposal as well: "Whoever stores waste or deposits, transits or otherwise disposes therewith and thus causes environmental damage or hazard shall be sentenced to imprisonment for up to two years" (Article 298.2). In such a case, gaining extensive profit by the act increases the term of imprisonment up to five years (298.4). The Code also provides protection for the animals by ruling that "whoever kills, destroys, processes, imports, exports, transits, handles, offers, mediates, obtains for him-/herself or for another a subject of an especially protected animal or herbal species shall be sentenced to imprisonment for up to three years" (Article 299.1), and "whoever maltreats an animal in an especially cruel or agonising manner shall be sentenced to imprisonment for up to two years" (Article 302.1). These are applicable to natural persons only. Fines for criminal offences can range from CZK 1,000 to CZK 2 million (EUR 40 to EUR 75,000) per day for up to 730 days. Courts can also prohibit an activity or halt participation in a public tender or concession procedure.

Act on Criminal Liability of Legal Entities and their prosecution[80] took effect as of January 1, 2012, introducing an entirely new procedure in the legal order of the Czech Republic. The act regulated the terms of criminal liability of legal entities, punishments, and protective measures that may be imposed on legal entities for the commitment of specified crimes, and the process of their prosecution. Thus, the Act extended the criminal sanctions for environmental violations to legal persons.[81]

Conclusive Remarks

It is argued in this chapter that regulatory EU policies such as the environmental policy are particularly prone to have administrative impacts. Concrete administrative implications can be observed in the member states via procedural obligations defined in the EU legislation. So, which administrative changes have been empirically validated in the context of the Czech institutional structure in environment?

Regarding the EU's environment assessment procedures, both the EIA and SEA procedures were adapted by the Czech Republic. EU's information procedures in environment, namely the freedom of access and public access to environmental information as well as the INSPIRE procedure were also adapted by the Czech Republic. An institutional framework was developed with an aim to clarify the division of work and their institutional responsibilities. Regarding the planning procedures, transposition of the EU directives gave way to the introduction of different planning procedures in environment in the Czech Republic in the last decade. EU's environmental auditing procedure, EMAS, was also adapted although it has not found a broad application in the Czech Republic. Finally, EU's sanctioning procedure in environment was transposed through two acts in the Czech Republic, introducing an entirely new procedure in the domestic legal order (see table 2.1).

CONCLUSION

To assess the domestic impact of the EU environmental policy, I distinguished two dimensions as administrative structures and administrative methods. I argued that on the one hand, EU policies put pressure on national administrations because of the structural requirements they imply. For instance, a regulation may call for the creation of new structures, such as an environmental agency. So, they imply a change in the presence and design of the domestic institutional structures. On the other hand, EU policies have an impact on the administrative methods, or the way of doing things, in the domestic contexts. Policy transfer results in emergence of new policy methods or change of the existing methods to be better suited to the obligations of EU policy.

In this chapter, I used these two dimensions as the empirical indicators of domestic institutional change. When analyzing these indicators, I posed the empirical question as: What are the institutional requirements implied by the EU's environmental policy and what are the Czech Republic's institutional responses to them?

Since the early 1990s the EU has made an increasing use of new policy methods as discussed in this chapter. These methods have triggered considerable adaptation pressures on the Czech Republic's domestic policy methods. Most of these measures had no base in Czech environmental policy structure before. Therefore, EU environmental policy methods clearly influenced the Czech administrative procedures. The empirical analysis in this chapter demonstrated that the Czech Republic adopted all these procedural measures, and thereby, changed its policy methods as a result of the impact of the EU policy methods.

The EU impact on the environmental policy methods in the Czech Republic is obvious, and indeed, it was recognizable even before the EU membership of the Czech Republic. Not only in the Czech Republic, but also in all Central

Table 2.1 Transfer of the EU's Environmental Policy Methods into Czech Republic

European Union		*Czech Republic*		
Policy Method	*Year of Adoption*	*Transfer of Policy Method*	*Year of Adoption*	*National Response*
Assessment Procedures (in environment)				
Environmental Impact Assessment	1985	Act on Environmental Impact Assessment	1992 + several amendments	Adaptation
Strategic Environmental Assessment	2001	Amendment to Act on Environmental Impact Assessment	2001	Adaptation
Information Procedures (in environment)				
Freedom of Access to Information on Environment	1990	Law on Access to Information on the Environment	1998	Adaptation
		Law on Free Access to Information	1999	
Public Access to Environmental Information	2003	Amendment to the Law on Access to Information on the Environment	2009	Adaptation
INSPIRE	2007			
Planning Procedures (in water and waste management)				
Plan for Water Supply and Sewerage Systems	1991	Water Supply and Sewerage Development Plan	2001	Adaptation
River Basin Management Plan	2000	Plan of Main River Basins of the Czech Republic	2001	Adaptation
Waste Management Plan	1975	Waste Management Plan	2003	Adaptation
Waste Prevention Plan	+ several amendments (latest in 2008)	Waste Prevention Plan	2013	Adaptation
Auditing Procedures (in environment)				
Eco-Management and Audit Scheme (EMAS)	1993; 2001; 2009	Eco-Management and Audit Scheme	1998; 2002	Adaptation
Sanctioning Procedures (in environment)				
Protection of Environment Through Criminal Law	2008	Criminal Code	2009	Adaptation
		Criminal Liability of Legal Entities and their Prosecution	2012	

Source: Author's own compilation

and Eastern European states, association agreements with the EU required some movement toward EU environmental standards. Their eventual road to EU membership required the full application of community legislation.

The empirical findings also validate a supporting research hypothesis. I argued before that the necessary condition for institutional change is a certain amount of misfit between the domestic policy procedures and the EU policy procedures. This supporting research hypothesis is applied to the administrative procedures in this chapter. For instance, regarding the administrative procedures in environment policy, an open and transparent form of administrative interest intermediation implied by the Environmental Information Directive was not compatible with the more closed practice to be found in the Czech Republic, where access to environmental information is generally restricted to parties directly affected by and involved in administrative activities and procedures. Such differences point out to the areas of misfit where we observe domestic change and adaptation to the EU policy methods.

However, the EU accession did not pose an empirically detectable effect upon the institutional structures of Czech environmental policy-making. As those structures largely came into being at the time of post-communist transition, and the EU environmental policy has evolved more or less in the same period, the Europeanization effect is less immediately distinguishable than it is for the older policy structures associated with more traditional domains such as foreign affairs, economy, or agriculture.

NOTES

1. Featherstone, K. (2003) "In the Name of Europe," in Featherstone, K. and Radaelli, C. (eds.), *The Politics of Europeanisation*, Oxford: Oxford University Press, p. 7.
2. Rometsch, D. and Wessels, W. (eds.) (1996) *The EU and Member States*, Manchester: Manchester University Press, p. xiii.
3. Ibid., p. 329.
4. Bulmer, S. and Burch, M. (1998) "Organising for Europe," *Public Administration* 76, p. 624.
5. DiMaggio and Powell (1991) "The Iron Cage Revisited: Institutional," p. 66.
6. Radaelli (2000), "Policy Transfer in the European Union: Institutional Isomorphism as a Source of Legitimacy," p. 27.
7. Dolowitz and Marsh (1996), "Who Learns What From Whom: A Review of the Policy Transfer Literature," p. 346.
8. Radaelli (2000), "Policy Transfer in the European Union: Institutional Isomorphism as a Source of Legitimacy," p. 28.
9. Knill and Lenschow (2001) "How Europe Matters," p. 125.
10. Bennett (1991) "What is Policy Convergence and What Causes It?" p. 220.

11. Ibid., p. 223.

12. Ibid., p. 223.

13. DiMaggio and Powell (1991) "The Iron Cage Revisited: Institutional," p. 67.

14. Radaelli (2000), "Policy Transfer in the European Union: Institutional Isomorphism as a Source of Legitimacy," p. 28.

15. Online Source: https://www.mzp.cz/en/ministry (Accessed on October 20, 2019).

16. Act. No. 114/1992 Gazette on the nature and landscape protection.

17. Online Source: http://www.ochranaprirody.cz/en/about-us/ (Accessed on October 20, 2019)

18. Based on the Article 54 of Act No. 219/2000 Coll. on the status of state institutions.

19. Online Source: https://www.cenia.cz/o-cenia/ (Accessed on October 20, 2019).

20. Ibid.

21. Act No. 282/1991 Coll., on the Czech Environmental Inspectorate.

22. Czech Environmental Inspectorate, Annual Report 2015, p. 6; see also online source: http://www.cizp.cz/Who-we-are (Accessed on October 20, 2019).

23. IMPEL, an international nonprofit association, has, as of January 2016, fifty members from thirty-five countries in Europe. Participants range from inspectors, law enforcement officers, policy-makers to managers. The core of the IMPEL activities concern awareness raising, capacity building, peer review, exchange of information and experiences on implementation, international enforcement collaboration as well as promoting and supporting the practicability and enforceability of European environmental legislation.

24. Act No. 388/1991 Coll.

25. State Environmental Fund of the Czech Republic, Annual Report 2015, p. 4; see also online source: https://www.sfzp.cz/ (Accessed on October 20, 2019).

26. Interview with the former Minister of Environment of the Czech Republic, Bedrich Moldan, February 27, 2020, Prague.

27. State Environmental Fund of the Czech Republic, Annual Report 2015, p. 19–20.

28. Ministerial Council Resolution of December 19, 1919.

29. MoE Decree No. 12/06 of December 12, 2006, on the issuance of the deed of foundation of a public research institution as amended by MoE Decree No. 2/11 of May 31, 2011.

30. T.G. Masaryk Water Research Institute, Annual Report 2018, p. 19.

31. Act No. 62/1988 on geological work, and MoE Decree No. 368/2004 on geological documentation.

32. Government Order of the Czechoslovak Republic No. 96/1953 Coll. of November 27, 1953, on the Institute of Hydrometeorology, with effect from January 1, 1954.

33. Measure No. 15/15 of the Ministry of the Environment of December 30, 2015, on the promulgation of the full wording of the Charter of the Czech Hydrometeorological Institute.

34. Online Source: http://www.clidata.cz/en/introduction/ (Accessed on October 20, 2019).

35. Government Resolution of July 30, 2003, No. 778.

36. Socio-Economic Development Committee; Strategic Management and Analysis Committee; Landscape, Water and Biodiversity Committee; Sustainable Energy Committee; Sustainable Municipalities Committee; Sustainable Transport Committee; Education for SD Committee; International Affairs Committee; and Committee on Sustainable Development Indicators.

37. Online Source: https://www.vlada.cz/en/ppov/council-for-sustainable-development/the-government-council-for-sustainable-development-of-the-czech-republic--153075/ (Accessed on October 20, 2019).

38. Constitutional Act No. 347/1997 Coll. of December 3, 1997, on the creation of higher territorial self-governing units and amending Constitutional Act of the Czech National Council No. 1/1993 Coll., Constitution of the Czech Republic.

39. In accordance with the law on regions, a "region" is a territorial community of citizens who have the right to self-government. At the same time, a "region" is a public corporation that has its own property that it administrates independently under the conditions stipulated by law and in accordance with its budget.

40. Interview with the former Minister of Environment of the Czech Republic, Bedrich Moldan, February 27, 2020, Prague.

41. Ibid.

42. Ibid.

43. Ibid.

44. Council Directive 85/337/EEC on the assessment of the effects of certain public and private projects on the environment, *Official Journal of the European Communities*, L 175/40. This Directive was amended on December 13, 2011, by the Directive 2011/92/EU. It was further amended on April 16, 2014, by Directive 2014/52/EU.

45. The EIA procedure can be summarized as follows: the developer may request the competent authority to say what should be covered by the EIA information to be provided by the developer (scoping stage); the developer must provide information on the environmental impact (EIA report); the environmental authorities and the public (and affected member states) must be informed and consulted; the competent authority decides, taken into consideration the results of consultations. The public is informed of the decision afterward and can challenge the decision before the courts.

46. Act No. 244/1992 Coll. on Environmental Impact Assessment.

47. Act No. 100/2001 Coll. of February 20, 2001, on Environmental Impact Assessment and on Amendments to Some Related Acts.

48. Specifically, the allegation was that the Czech Republic failed to comply with its obligations under Article 3, paragraph 9, and Articles 6 and 9 of the Aarhus Convention; so that the members of the public in other countries (in this case, Germany) did not have the same possibility to participate in the decision-making procedure (concerning the Temelín Nuclear Power Plant as members of the public in Czech Republic). For more information: Aarhus Convention Compliance Committee ACCC/C/2012/71 dated May 31, 2012. See also United Nations Economic and Social Council ECE/MP.PP/C.1/2017/3 dated December 29, 2016.

49. Act No. 39/2015 of February 10, 2015, amending Act no. 100/2001 Coll., on Environmental Impact Assessment and on Amendment to certain related acts.

50. Act No. 326/2017 of September 5, 2017, amending Act No. 100/2001 Coll., on Environmental Impact Assessment and on Amendment to Certain Related Acts.

51. Council Directive 2001/42/EC on the assessment of the effects of certain plans and programmes on the environment, *Official Journal of the European Communities*, L 197/30.

52. The SEA procedure can be summarized as follows: an environmental report is prepared in which the likely significant effects on the environment and the reasonable alternatives of the proposed plan or program are identified. The public and the environmental authorities are informed and consulted on the draft plan or program and the environmental report prepared. As regards plans and programs which are likely to have significant effects on the environment in another member state, the member state in whose territory the plan or program is being prepared must consult the other member state(s). The environmental report and the results of the consultations are taken into account before adoption. Once the plan or program is adopted, the environmental authorities and the public are informed and relevant information is made available to them.

53. Act No. 100/2001 Coll. of February 20, 2001, on Environmental Impact Assessment and on Amendments to Some Related Acts.

54. Council Directive 90/313/EEC on the freedom of access to information on the environment, *Official Journal of the European Communities*, L 158/56.

55. Law No. 123/1998 of May 13, 1998, on Access to Information on the Environment.

56. Law on Free Access to Information, dated May 11, 1999, Act No. 106/1999 Coll.

57. Council Directive 2003/4/EC of January 28, 2003, on Public Access to Environmental Information, *Official Journal of the European Communities*, L 41/26.

58. Act No. 365/2000 on public administration information systems.

59. Online Source: http://www.geology.cz/extranet-eng/sgs (Accessed on February 20, 2020).

60. Council Directive 2007/2/EC of March 14, 2007, establishing an Infrastructure for Spatial Information in the European Community, *Official Journal of the European Union*, L 108/1.

61. Act No. 380/2009.

62. Council Directive 2000/60/EC of October 23, 2000, establishing a framework for Community action in the field of water policy, *Official Journal of the European Communities*, L 327/1.

63. Act No. 254/2001 Coll. on Water and Amendments to Some Acts.

64. Government Resolution No.562.

65. Council Directive 91/271/EEC of May 21, 1991, concerning urban waste water treatment, *Official Journal of the European Communities*, L 135/40.

66. Act No. 274/2001 Coll. of July 10, 2001, on Water Supply and Sewerage Systems.

67. Directive 75/442/EEC of July 15, 1975, on waste, *Official Journal of the European Communities*, L 194/39.

68. Directive 2008/98/EC of November 19, 2008, on waste and repealing certain directives, *Official Journal of the European Communities*, L 312/4.

69. Act No. 185/2001 Coll. of May 15, 2001, on waste and amending some other laws.

70. The binding part of the Plan was announced by Government Order No.197/2003 and amended by the Government Order No.473/2009 Coll. The Government Order No.181/2013 Coll. extended the Plan until December 31, 2014.

71. According to the Waste Framework Directive, Waste Prevention Programs can either be included in waste management plans or may be developed separately. Act No. 154/2010 Coll. incorporated the waste prevention program into the waste management plan of the Czech Republic.

72. Government Resolution No. 869/2014.

73. Council Regulation No.1836/1993 of June 29, 1993, on allowing voluntary participation by companies in the industrial sector in a Community eco-management and audit scheme, *Official Journal of the European Communities*, L 168/1.

74. Resolution of the Government of the Czech Republic No. 466/1998 of July 1, 1998.

75. Hyrslova, J. and Hajek, M. (2005) "Environmental Management Accounting in the Framework of EMAS II in the Czech Republic," in Rikhardsson, P.M., Bennett, M., Bouma, J.J., Schaltegger, S. (eds.), *Implementing Environmental Management Accounting*, Dordrecht: Springer, p. 284.

76. The updated version of the EMAS program was based on Regulation No.761/2001 of the European Parliament and of the Council (EMAS II). This was adopted by the Czech Republic by the Government Resolution No. 651 of June 19, 2002, on the updated EMAS program.

77. Regulation No. 1221/2009 of November 25, 2009, on the voluntary participation by organizations in a Community EMAS, *Official Journal of the European Union*, L 342/1.

78. Council Directive 2008/99/EC of November 19, 2008, on the protection of the environment through criminal law, *Official Journal of the European Union*, L 328/ 28.

79. The term "larger area" in these articles was defined as "territory of total area of at least three hectare, and in case of surface waters in water bodies at least one hectare of water surface, and in case of stream at least two kilometres of its length" (Article 296.1).

80. Act No. 418/2011 Coll. of October 27, 2011.

81. In the context of the criminal sanctions for the environmental violations, the Act stated the bases of criminality for legal entities in Article 7 as "damage and threat to the environment (Section 293), damage and threat to the environment due to negligence (Section 294), water source damage (Section 294a and 295), unauthorized discharge of pollutants (Section 297), unauthorized waste management (Section 298), unauthorized production and other handling of ozone depleting substances (Section 298a), unauthorized management of protected wildlife fauna and flora (Section 299), unauthorized management of protected wild fauna and flora negligently (Section 300), damage to protected parts of nature" (Section 301).

Chapter 3

Water Management

INTRODUCTION: IMPORTANCE OF THE WATER MANAGEMENT POLICY

Water is essential for life, it is an indispensable resource for the economy, and it also plays a fundamental role in the climate regulation cycle. The management and protection of water resources,[1] of fresh and salt water ecosystems and of the drinking water, is, therefore, important cornerstones of the environmental protection.

Major watercourses in the Czech Republic are Elbe River (370 km) with Vltava River (431 km) and Ohře River (254 km) in Bohemia; Morava River (269 km) with Dyje River (194 km) in South Moravia; and Oder River (135 km) with Opava River (131 km) in the north of Moravia and in Silesia. Because the country's almost all major watercourses drain water into neighboring countries, fresh water resources of the Czech Republic are entirely dependent on atmospheric precipitation. However, due to hydrological extremes, water regime is very fluctuating, and drought and flood situations become two extreme conditions for the water management policy to deal with. In the Czech Republic, hydrological drought has persisted since 2014, which is why more consideration is placed on this phenomenon with an aim to ensure sufficient amount of drinking water as well as water for industry and agriculture in adequate quality. On the other hand, the occurrence of floods in the Czech Republic have increased since 1997, and the country has been affected by a number of major to extreme floods, accompanied by damage to property and loss of human lives.

The groundwater represents an important link of the hydrological cycle through the maintenance of wetlands and river flows, acting as a buffer through dry periods. It provides the base flow for surface water systems, that is, the water which feeds rivers all year round; and many of them are used for water supply and recreation. For this reason, deterioration of groundwater quality and quantity directly affects other related aquatic and terrestrial ecosystems. The pollution from domestic, agricultural, and industrial sources is a major concern for the groundwater quality.

The wastewater is another important determinant in this area. The quantity of wastewater and the pollution it discharged into surface water directly affect the ecosystems of the aquatic environment. The most important components of wastewater pollution are the organic compounds, nutrients (especially phosphorus and nitrogen), and hazardous substances. Excess nutrients can create eutrophication, a process characterized by increased plant growth, problematic algal blooms, depletion of oxygen, loss of life in bottom water, and undesirable disturbance of the balance of organisms present in the water. The polluted water not only damages the environment, but also can be a source of infectious diseases for human such as viral hepatitis A, dysentery, and salmonella. Hazardous substances pollute drinking water sources (especially groundwater), and get into plant and animal tissues, and then into the food chain of animals and humans, where they occur a long time after their release. Briefly, water pollution through hazardous substances and chemicals threatens both aquatic ecosystems and human health.

The reduction of water pollution is an essential element of protecting the water ecosystems, and therefore, maintaining a sustainable environment and keeping up economic and social development. For this reason, water management policy is significant for all countries. This policy area links the issues of precipitation, water quality, sources of pollution, water processing, and watercourse administration. The crucial impact, in this context, comes from the European legislation transposed into the national legislation and to policy strategies.

EUROPEAN AND DOMESTIC POLICY NORMS IN WATER MANAGEMENT

EU requirements have been an important driver of improvement in water management and water quality in the EU member states. The EU Strategy Papers and Programmes established community-wide objectives in water management area, and they have been the initiator and driver of changes and developments in the water management policies of the member states. In the first part of this analysis, I will analyze the European Union's and the Czech Republic's *executive portfolio in water management* with an aim to find out the level of resonance of general policy norms, objectives, and strategies.

In 2001, the European Council adopted the first EU Sustainable Development Strategy in Goteborg. The presidency conclusions declared that "sustainable development is a fundamental objective under the Treaties. That requires dealing with economic, social, and environmental policies in a mutually reinforcing way. Failure to reverse trends that threaten future quality of life will steeply increase the costs to society or make those trends irreversible."[2]

In 2002, the Sixth Community Environment Action Programme was established covering a period of ten years.[3] It addressed the key environmental objectives and priorities based on an assessment of the environment and of prevailing trends that require a lead from the Community. The program promoted the integration of environmental concerns in all Community policies and achievement of sustainable development throughout the Community (Article 1.1). The key environmental priorities were defined as climate change, nature and biodiversity, environment and health and quality of life, natural resources and wastes (Article 1.4). Regarding the sustainable use and quality of water, priority actions were defined as "ensuring a high level of protection of surface and groundwater, preventing pollution and promoting sustainable water use; ensuring full implementation of the Water Framework Directive, aiming at a good ecological, chemical and quantitative water status and a coherent and sustainable water management" (Article 2.7.e). To attain these objectives, the program stated that development of new Community legislation and more effective implementation and enforcement of Community legislation are required. This policy strategy is to be pursued via (a) increased measures to improve respect for Community rules on the protection of the environment and addressing infringements of environmental legislation; (b) promotion of improved standards of permitting, inspection, monitoring, and enforcement by member states; (c) more systematic review of the application of environmental legislation across the member states; and (d) improved exchange of information on the best practice on implementation (Article 3).

In 2006, the Renewed EU Sustainable Development Strategy was adopted. This strategy paper aimed to present a single, coherent strategy on how the EU will more effectively meet its commitment to the principles of sustainable development. The strategy paper identified its operational objectives in the area of conservation and management of natural resources as "(a) improving management of natural resources and avoiding over-exploitation of renewable natural resources such as fisheries, biodiversity, water, air, soil and atmosphere, and (b) ensuring that member states work towards improving integrated water resources management."[4] The sustainable use of natural resources, particularly water resources, has been a key priority and challenge in Europe for many decades, especially considering the impacts of the climate change on the ecosystems. Changing water levels and temperatures, as well as water quality affect food production and ecosystems and their related services on which society depends. Thus, the challenge for European policy-makers is to understand these climate change impacts and to develop and implement policies to ensure an optimal level of adaptation. These adaptation policies and measures require solidarity among EU member states to ensure that timely and effective adaptation measures are taken, and coherency is ensured across different sectors and levels of governance. With this aim, in 2009, the EU adopted a White Paper titled "Adapting to Climate Change: Towards a European Framework for Action"[5] with an aim to develop strategies for the management and conservation of water. The strategy was to improve the EU's resilience to deal with the impact of climate change in two phases: "Phase 1 (2009-2012) focused on building a solid knowledge base on the impact and consequences of the climate change for the EU, integrating adaptation into key EU policy areas, employing both market-based and public-private partnership instruments, and promoting international cooperation on adaptation. These strategies are to be implemented during Phase 2, which commences in 2013."[6] The White Paper proposed three actions for sustainable water management: (1) develop guidelines and a set of tools by the end of 2009 to ensure that the river basin management plans (RBMPs) are climate-proofed; (2) assess the need for further measures to enhance water efficiency in agriculture, households, and buildings; and (3) explore the potential for policies and measures to boost ecosystem storage capacity for water.[7]

These policy objectives and strategies were further developed in the Blueprint to Safeguard Europe's Water Resources[8] which set out key actions that need to be taken by water managers and policy-makers to address the challenges faced by the aquatic environment in Europe in order to secure the availability of good-quality water for sustainable and equitable water use. The strategy paper identified persisting problem areas in relation to (1) land use/ecological status of EU waters; (2) chemical status and pollution of EU waters; (3) water efficiency; (4) vulnerability of EU waters; and (5) crosscutting issues. Although there have been significant improvements in the chemical quality of water bodies owing to the implementation of relevant EU directives in the member states, the situation as regards the priority substances was "below expectations": "diffuse and point-source pollution are still significant pressures on water environment, and eutrophication due to excessive nutrient load remain a major threat to the good status of waters."[9] The proposed actions to counter these threats were to extend nitrate vulnerable zones and reinforce action programs, and to improve compliance rates on wastewater treatment through long-term investment planning and implementation plans.[10] Establishment of a common knowledge base for all member states was proposed to implement these policy strategies. Thus, Water Information System for Europe (WISE) was established to gather EU-wide information on water status and policy. Besides, Innovation Partnerships on Water[11] and Agricultural Productivity and Sustainability[12] were launched in 2012 with an aim to find solutions to water challenges in urban, industrial, and agriculture contexts.

The Fourth European Water Conference, organized by the European Commission's Environment Directorate-General, was held in Brussels on March 23–24, 2015. The conference focused on four areas: experiences from the first cycle of implementation of the WFD; achievements and lessons learned; preparations and expectations for the second RBMPs; and the financial support for such initiatives that are available from the EU in the new financing period 2014–2020. During the conference, policy-makers stressed the need for improved water quality and better quantitative management of water, and underlined that progress in some areas had not gone far enough or had not been delivered at a rapid enough pace. The conference also underlined increased flood and drought risks that may be associated with climate change, and the need to develop green infrastructure such as water retention measures, alongside more traditional engineering solutions.

The Seventh Environment Action Programme, which entered into force in January 2014, was the guiding European environment policy strategy until 2020. It set out the long-term objectives regarding the sustainable use of resources, and identified three key objectives for this purpose: "protect, conserve and enhance the Union's natural capital; turn the Union into a resource-efficient, green, and competitive low-carbon economy; and reduce threats to human health and well-being linked to pollution, chemical substances, and the impacts of climate change." To get there, the EAP expressed the commitment of the EU, national authorities, and stakeholders to speed up the delivery of the objectives stated in the Blueprint to Safeguard Europe's Water Resources. These policy goals are to be fulfilled through "an enabling framework" that has four priority objectives: "better implementation of legislation, better information by improving the knowledge base, more and wiser investment for the environment, and full integration of environmental requirements and considerations into other policies."[13] The member states were expected to ensure that these measures are implemented and priority objectives are met by 2020.

Europe 2020: A Strategy for Smart, Sustainable and Inclusive Growth[14] put forward three mutually reinforcing priorities: smart growth (developing an economy based on knowledge and innovation); sustainable growth (promoting a more resource-efficient, greener, and more competitive economy); and inclusive growth (fostering a high-employment economy delivering social and territorial cohesion). In the area of sustainable growth, the policy objective was to establish a vision of structural and technological changes required to move to a low-carbon, resource-efficient, and climate-resilient economy.

To attain these objectives, the policy strategies were defined as preventing environmental degradation and biodiversity loss, ensuring disaster prevention and response, addressing climate change through adaptation measures based on more efficient use of resources, and mobilizing EU financial instruments as part of a consistent funding strategy. In order to ensure timely and effective implementation, Europe 2020 required action at both EU level and member states level. In this context, each member state was required to draw up a Stability and Convergence Programme and a National Reform Programme, in which they define their exit strategies and describe how to implement them; report on their progress toward these targets; present their key structural reforms; and identify their bottlenecks to growth. These programs should be submitted to the Commission for assessment so that the Commission can prepare and publish a yearly report on the progress toward meeting the agreed headline targets.

After becoming a member of the EU, the Czech Republic adopted these aforementioned water policy norms and strategies. We will start the analysis of this adaptation with an important policy paper: "Strategic Framework for Sustainable Development of the Czech Republic" (2010). The policy paper presented the main threats for the security of water resources as "pollution with municipal and industrial wastewater, large-scale drainage of land, intensification of agriculture as part of large-scale production, and changed water balance and floods due to the climate change."[15] The policy paper stressed that "intensive farming in the Czech Republic is disturbing the landscape water regime and causing water pollution by means of contamination with foreign substances from fertilizers and pesticides."[16] Another serious concern was the more frequent occurrence of alternating extreme weather events (droughts and floods). Floods pose the greatest direct risk for the Czech Republic due to creating substantial material damage, health damage, and ecological damage. Equally serious was the risk of increasing temperatures caused by the drying of surface and subsurface water, the inability of excessively dry soil to absorb greater amounts of short-term rainfall and, in consequence, the rapid runoff. With an aim to counter such threats, the policy paper presented its policy objectives as: (1) ensuring sufficient quantity and quality of surface water and groundwater; (2) improving the landscape water regime and improving the overall ecological, hydrological, and hydromorphological condition of aquatic ecosystems; (3) promoting sound and close-to-nature farming methods; and (4) ensuring readiness to deal with emergencies related to climate change.

State Environmental Policy of the Czech Republic (2012–2020) set the framework for an effective protection of environment in the Czech Republic. The policy paper stated that the most important and urgent problems to be tackled in the short and medium-term period were "urban wastewater treatment, improvement of drinking water supply, achieving good ecological condition of waters, and avoiding impacts of natural disasters."[17] In this context, the state environmental policy defined its primary objective as "the protection and sustainable use of resources based on the legislative framework put forward by the EU Water Framework Directive (WFD) and on objectives covered in other water directives in the field of urban wastewater treatment, nitrates, bathing water and others."[18] The policy strategies were

> implementation of the monitoring programmes for the surface and groundwater to find out all measures under the WFD are implemented; achievement of at least a good ecological status and good chemical status of bodies of surface water; achievement of good chemical and quantitative status of bodies of groundwater; ensuring protection of water in protected territories defined in accordance with the WFD.[19]

In the interview with the first Minister of Environment of the Czech Republic, Bedrich Moldan expressed that the main problem associated with these earlier policy papers as that "they were too general and not specifically translated into concrete steps and concrete policies. Therefore, they were not practically implemented well."[20] He observed another problem that "these strategy papers did not reach a wide population; only Ministerial staff and policy-makers were aware of them. In the early years of the policy development, this signified a gap between the Ministry and the practitioners."[21]

Strategy on Adaptation to Climate Change in the Czech Republic was prepared in 2009 under an interministerial cooperation[22] with the Ministry of Environment as the coordinator of the preparation of the whole document. The strategy paper was based on the White Paper of the European Commission, "Adapting to Climate Change: Towards a European Framework for Action" (2009). The strategy paper for the Czech Republic was prepared for the years 2015–2020 with a view to 2030; and identified the priority areas that were expected to be the most affected by climate change. By doing so, it provided a structured overview of the risks and anticipated impacts of climate change in the priority areas, defined general principles of adaptation measures, pointed out intersectoral linkages, and provided guidelines for and examples of suitable adaptation measures.[23]

The implementing document of the Adaptation Strategy is the National Action Plan on Adaptation to Climate Change. The National Action Plan was prepared in 2016 as a product of interdepartmental cooperation, with the Ministry of the Environment as the coordinator, and was approved by Government Resolution No. 34 of January 16, 2017. The Action Plan is structured in accordance with the manifestations of the climate change, such as long-term drought and floods, increasing temperature, extreme meteorological phenomena, and wildfires. The Action Plan elaborated the measures included in the Adaptation Strategy into specific tasks within thirty-three specific objectives and one crosscutting objective, and assigned them to relevant government bodies with specific responsibilities, implementation deadlines, and measures to be taken against the climate change impacts.

Strategic Framework Czech Republic 2030[24] was adopted in 2017 and replaced the Strategic Framework of Sustainable Development (2010). Based on the structural analysis of the contemporary conditions and trends, the new document formulated strategic objectives that the Czech Republic should fulfill by the year 2030. The former Minister of Environment, Bedrich Moldan described the plan as "a good document, meaning that it was better than the previous plans. However, it was still not clear how it will be implemented and how it will have an impact in practice."[25] The Report observed that the ecosystems of the Czech Republic are under significant pressure from intensive agriculture and forestry which contribute to low quality of a significant part of surface waters and their excessive enrichment with nutrients, and fast water discharge from landscape. Therefore, reversing these trends by 2030 is necessary by slowing down drainage water from the landscape and improving the quality of surface water and groundwater. To attain these objectives, the policy strategy was defined as restoring riverbeds, river floodplains, wetlands and other water reservoirs by long-term measures. With regard to pesticides, the objective was to reduce the number of surface water bodies that have exceeded environmental quality limits due to the presence of pesticide residues. Organic farming and more precise use of pesticides were also encouraged with an aim to achieve these objectives.

Conclusive Remarks

In the previous section, I put forward that the EU-accession process produces two varieties of policy transfer: hard transfer of programs and their implementation, and soft transfer of ideas, concepts, and attitudes. In this section, I present that after its accession to the EU, the Czech Republic transferred the ideas, concepts, and strategies presented in the EU's water management policy documents. This means that EU policy norms and policy objectives have been adopted by the Czech Republic as part of its membership obligations. As a result, national policy norms and strategies were formulated in accordance with the EU strategic documents. A comparative analysis of policy norms and objectives set by the EU and by the Czech Republic in the last two decades demonstrates and validates the theoretical hypothesis of the soft transfer of ideas, concepts, and attitudes from the supranational level to the national level. Next section analyzes the hard transfer of the policy legislation.

EUROPEAN AND DOMESTIC POLICY LEGISLATION IN WATER MANAGEMENT

This section provides an overview of the transfer of the EU water management legislation into the domestic structure of the Czech Republic. At the European level, the Water Framework Directive and other related directives established community-wide objectives to reduce water pollution, promote better use of water and protection of water, and mitigate the impact of floods and droughts. These Directives have been transposed into member states' legislation in variable degrees. In this section, I will focus on the development and amendment of the domestic water management legislation in the Czech Republic as a response to the EU-level legislation.

Important legislative acts on water management at the EU level include the Water Framework Directive, Nitrates Directive, Urban Waste Water Treatment Directive, Ground Water Directive, and Drinking Water Directive. This section describes the objectives and requirements of these main EU directives, and investigates their adoption and transposition into the legal framework of the Czech Republic.

I start this legislative analysis with the analysis of the primary legislation in this policy field, Council Directive 2000/60/EC, the Water Framework Directive, which was adopted on October 23, 2000. The WFD established a framework for water management aiming at sustainable water use, improvement of water quality, progressive reduction of pollution of groundwater, and enhanced protection and improvement of the aquatic environment in the EU (Article 1). It stated that these environmental objectives are to be achieved by the member states for surface waters (Article 4.a), for groundwaters (Article 4.b), and for protected areas (Article 4.c). For this purpose, it required member states to bring into force the laws, regulations, and administrative provisions by December 22, 2003 (Article 24).

The policy strategies included in the WFD can be analyzed in three categories:

On river basin management, each member state was required to establish for each river basin district, or for the part of an international river basin district within its territory, a program of measures (Article 11), which include the basic measures (Article 11.3), and where necessary, supplementary measures (Article 11.4). The program of measures shall be established at the latest nine years after the date of entry into force of this directive, and all the measures shall be made operational twelve years after that date at the latest (Article 11.7). These programs should be reviewed and, if necessary, updated every six years thereafter (Article 11.8). Member states shall produce a river basin management plan for each river basin district lying entirely within their territory (Article 13.1), as well as a single international river basin management plan for those international river basin districts falling within the Community (Article 13.2). These plans should be sent to the Commission, and to any other member state concerned, within three months of their publication (Article 15).

On strategies against pollution of surface water, the WFD assigned the European Parliament and the Council to adopt specific measures against pollution of water by individual pollutants or by groups of pollutants presenting a significant risk to the aquatic environment and to the drinking water (Article 16.1). In this context, the Commission was assigned to submit a proposal setting out a list of priority substances (Article 16.2) including priority hazardous substances (Article 16.3). This list should be reviewed every four years (Article 16.4). The Commission was also assigned the task of setting quality standards applicable to the concentrations of the priority substances in the surface water (Article 16.7).

On strategies to prevent and control pollution of groundwater, the WFD assigned the European Parliament and the Council to adopt specific measures to prevent and control groundwater pollution with an aim to achieve the objective of good groundwater chemical status (Article 17.1).

The WFD was transposed into the Czech Republic by Water Act No.254/2001 of June 28, 2001. The Water Act is the main national document in the water management area that incorporated the requirements of European directives. The purposes of the Act were delineated as "to protect surface water and groundwater; stipulate conditions for economic utilisation of water resources whilst preserving and improving the quality of surface water and groundwater; create conditions for reducing the adverse effects of floods and drought; and ensure the safety of water management structures in accordance with the European Community Law" (Section 1.1). The Act became effective on January 1, 2002.

The policy strategies established by the Water Act can be analyzed in three categories:

The requirements of the Article 13.1 of the WFD on *river basin management* were incorporated by the Water Act. According to the Act, the Plan of Main River Basins of the Czech Republic is to be prepared as a strategic water planning document (Section 24.1) by the Ministry of Agriculture in cooperation with the Ministry of Environment (Section 24.2). The Act also stated that this plan shall be "fundamental to the execution of the state administration for regional land use planning, regional decision-making, and granting of construction permits" (Section 23.2). In accordance with the WFD, the measures for water protection must include the principle measures and, where necessary, supplementary measures (Section 26.2) and they must be implemented within three years from approval of the Plan (Section 26.1).

On protected areas, the WFD stated that "Member States shall ensure the establishment of a register of all areas lying within each river basin district which require special protection under specific Community legislation for the protection of their surface water and groundwater (Article 6), including the water intended for human consumption providing more than 10 m^3 a day as an average" (Article 7.1). The Water Act incorporated this requirement into national legislation by stipulating that "to ensure protection of the yield, quality and wholesomeness of groundwater and surface water resources used or usable for drinking water supply with an average withdrawal of more than 10,000 m^3 per year, the water authority shall determine protected zones" (Section 30.1).

On harmful substances, the Water Act stated that "whoever works with harmful substances must take adequate precautions to prevent their entry into surface water or groundwater and to prevent them from endangering the environment" (Section 39.1). It provided the list of harmful substances in Annex I. According to the Act, "the monitoring and assessment of the status of surface water and groundwater and the operation of the public administration information systems are carried out by river basin administrators and other professional institutions authorized by the MoA or by the MoE" (Section 21.4).

In accordance with the Water Act Section 24, the Plan of Main River Basins of the Czech Republic was prepared by the MoA in cooperation with the MoE, and approved by the Resolution No. 562 of the government on May 23, 2007. The first RBMP was prepared in compliance with the requirements of the WFD and covered six-year period from 2009 until 2015. The Plan presented policy objectives, measures to be implemented, and strategy for their funding in three main areas: protection of water, flood protection, and water services.[26] In its evaluation report in 2012, the European Commission stated that "the first RBMP presented significant gaps regarding the monitoring system, assessment of pressures and classification methodology for water bodies' status, resulting in a high level of uncertainty."[27]

The second RBMP was prepared for the term 2016–2021, and was approved in 2015. In accordance with Article 15 of the WFD, it was necessary to ensure the duty to submit a report to the EU within three months following the completion and approval procedure. Therefore, the Czech Republic reported the second generation of RBMP to the European Commission, and the Commission presented its evaluations in 2019. In its evaluation report, the European Commission stated that "there are gaps in the apportionment of pressures, as no information was found in the RBMPs on which activities/sectors are contributing significantly to the failure of good chemical status in surface waters and to the different impacts causing failure of good ecological status/potential for both surface and groundwater bodies."[28] Concerning the achievement of measures so far, it noted that "a large number of measures from the first RBMPs were not implemented due to lack of funds, land property issues, technical, legal or natural conditions."[29] Another weakness was presented as that "the level of ambition and expected progress on the programme of measures for the second cycle is low with small reductions expected at the end of the second cycle."[30] The evaluation report concluded that the Czech Republic should "set objectives for different types of protected areas, expand monitoring sites to cover all protected areas, and demonstrate a clear financial commitment for the implementation of measures."[31]

Another important legislative act at the EU level is the Nitrates Directive. The reason for setting up a common EU legislation in this area was that nitrate concentrations in EU waters were rising and in some areas they were reaching values that are disproportionate to the requirements for drinking water quality. This trend was not consistent with the EU's goal of environmental sustainability. With this concern, the Council adopted *Directive 91/676/EEC, the Nitrates Directive,* on December 12, 1991. This Directive is one of the key instruments in the protection of waters of the EU against agricultural pressures. Indeed, the main objective of the Directive was defined as "to reduce water pollution caused or induced by nitrates from agricultural sources and prevent further pollution" (Article 1).

The Nitrates Directive dealt generally with all groundwater and surface waters polluted by nitrates or endangered by nitrates pollution as well as the marine and coastal waters where the nitrogen-rich runoff from the land facilitates eutrophication. The Directive required all member states to identify water pollution by nitrates from agricultural sources, and designate vulnerable zones in their territories within a two-year period following the notification of this Directive (Article 3.2). The member states were required to implement measures to reduce leakage of nitrates from agricultural sources into such vulnerable zones by action plans (Article 5.1) and implement these action plans within four years of their establishment (Article 5.4). The Directive also required member states to review these action plans every four years taking into account changes and factors unforeseen at the time of the previous designation (Article 3.4).

This Directive was transposed in the Czech Republic by five national regulations:

Act No. 156/1998[32] (Act on Fertilisers) regulated in accordance with European Community law, "the conditions of placing into circulation, storage and usage of fertilisers, supplementary soil substances, supplementary plant substances and substrates, the conditions of agrochemical testing of agriculture soil, the conditions of soil properties determination of forest land and conditions of usage of treated sludges" (Section 1). The Act authorized the Central Institute for Supervising and Testing in Agriculture to decide on fertilizer registration upon the application of producer and regulated the registration procedure (Sections 4–6).

Act No. 254/2001 (Water Act) required the government to specify by decree vulnerable zones and regulate the use and storage of fertilizers and manure, crop rotation, and implementation of erosion-control measures. The action program and specification of vulnerable zones is subject to review and possible amendment at regular intervals not exceeding four years (Article 33.2).

The first phase of implementation of the Nitrates Directive in the Czech Republic was completed in 2003 with the adoption of Government Regulation No. 103/2003.[33] The Regulation designated an action plan which applies to natural or legal persons who carry out agricultural production in vulnerable areas (Article 4). The proposed action plan regulated the period of prohibition of fertilization (Article 6), use of nitrogen fertilizers according to the soil and climatic conditions of the habitat (Article 7), restrictions on the use of organic nitrogen (Article 8), storage of nitrogen fertilizers in vulnerable areas (Article 9), and management of the agricultural land adjacent to surface water bodies (Article 12). The list of vulnerable areas was provided and it was specified that the review of the delimitation of vulnerable areas shall be carried out by a professional body authorized by the MoE (Article 3.1) and this review shall be submitted by the MoE to the government no later than four years from the date of entry into force of this Regulation (Article 3.4).

At the end of the first phase, Government Regulation No. 219/2007 amended the Regulation 103/2003, and provided a revised list of vulnerable zones and measures in these areas. The Action Program was revised at the end of 2007 and in April 2008 the amendments were announced by the Government Regulation No. 108/2008.

The vulnerable zones were revised at the end of the second phase of implementation of the Nitrates Directive. Government Regulation No. 262/2012 declared the reviewed list of vulnerable areas (Article 1). The review was carried out by the MoE on the basis of identification of surface or groundwater polluted or endangered by nitrates from agricultural sources (Article 3). The Regulation presented an Action Programme which specified the measures on the restriction of organic nitrogen use (Article 8), measures on the use (Article 7) and storage (Article 9) of nitrogenous fertilizers, crop rotation (Article 10) and no fertilization period (Article 6). It took effect on January 1, 2014.

This comparative analysis shows that the EU legislation on setting up vulnerable zones and stricter measures for those areas were transposed into the Czech Republic by several acts and regulations, and the requirement of the EU legislation was fulfilled by designing vulnerable zones, preparing an action program, and setting up measures.

Council Directive 91/271/EEC, the Urban Waste Water Treatment Directive (UWWD), was adopted on May 21, 1991, concerning the collection, treatment, and discharge of urban wastewater. The objective of the Directive was to protect the water environment from the adverse effects of the wastewater discharges (Article 1). The UWWD required member states to ensure that all agglomerations over 2,000 population are provided with collecting systems for urban wastewater by designated deadlines (Article 3.1) and that urban wastewater entering collecting systems shall be subject to secondary treatment or an equal treatment for agglomerations over 2,000 population before discharge (Article 4.1). The UWWD stipulated that the discharge of industrial wastewater into collecting systems and urban wastewater treatment plants is subject to prior regulations and/or specific authorizations by the competent authority (Article 11.1). Member states shall identify sensitive areas according to the criteria laid down in Annex of this Directive (Article 5.1) and review the identification of sensitive areas at intervals of no more than four years (Article 5.6). It also required member states to establish a program for the implementation of this Directive by December 31, 1993 (Article 17.1) and provide the Commission by June 30 every two years with an update of the information on the program (Article 17.3).

In the Czech Republic, various governmental responsibilities pertaining to water supply and sewerage sectors are divided among these institutions: MoE, MoA, Ministry of Finance, Ministry of Health, Ministry for Regional Development, counties, and municipalities. The requirements of the UWWD were transposed into the Czech Republic by these legislative acts:

Act No. 254/2001 (Water Act) defined the rights and obligations of all entities whose activities may affect the quantity or quality of water. The Act stated that any person using surface water or groundwater is obliged to protect these waters and to ensure their economical and efficient utilization pursuant to the conditions of this Act (Section 5.1). It also regulated the permission for water use (Sections 8–13) and the permission for discharge of wastewater containing especially harmful substances into sewerage systems (Section 16).

The Water Act assigned the government to establish acceptable values for pollution indicators in the wastewater discharged into watercourses (Section 38.5). Upon this assignment, the government established these values in Government Regulation No. 61/2003[34] on indicators and values of acceptable pollution of surface water and wastewater. This Regulation became valid on March 1, 2003. While Directive 91/271/EEC establishes strict values for each pollution indicator,[35] Government Order No. 61/2003 establishes a range defined between an acceptable amount and maximum allowed amount.[36]

This Regulation was amended by Government Regulation No. 401/2015[37] on indicators and values of permissible pollution of surface waters and wastewaters. This Regulation dealt with the setting of wastewater pollution limits for discharges from a wastewater treatment plant and stipulated that the sensitive areas be monitored for the levels of nitrogen and phosphorus.

Act No. 274/2001 (Act on Water Supply and Sewerage Systems) aimed to regulate the relations arising from the development, construction, and operation of water supply and sewerage systems as well as the competence of bodies of territorial self-governing units and administrative authorities in this area (Article 1.1). In this context, the law regulated the rights and obligations of the builder, owner, and operator in the construction, maintenance, and operation of sewerage (Articles 7–9), management of crisis situations (Article 21), water supply and sewerage network protection (Article 23), and the role of public administration in these issues (Articles 25–30).

As mentioned earlier, Article 17 of the UWWD required member states to adopt national and regional development plans for water supply and sewerage systems. The Act No. 274/2001 gave responsibility to regional units to independently prepare and approve the water supply and sewerage development plan for their territory including the drinking water supply, sewage, and wastewater treatment in the region. These development plans must be economical and contain the most technically appropriate solutions and links to the development plan for the neighboring regions (Article 4.1). The MoA was the main authority to process, update, and approve these development plans (Article 4.7). Regarding the water supply and sewerage development plan for the state, the MoA shall ensure the elaboration and updating of this plan (Article 29.1.b). Pursuant to this article, the MoA prepared the Development Plan for Water Supply and Sewerage Systems. The Plan included the standpoints of MoA and the updates of the Regional Development Plans for Water Supply and Sewerage Systems. The policy strategies were delineated as the construction and reconstruction of waste water treatment plants (WWTP) in communities over population of 2,000 and increasing the proportion of inhabitants connected to public sewerage systems and to the public sewerage systems connected to a WWTP. The plan was perceived by key policy-makers as "a concrete and important step for the implementation of water management policy objectives."[38]

Another important EU directive is the Council Directive 2006/118/EC, the Ground Water Directive. This Directive complemented the WFD by setting groundwater quality standards and introducing measures to prevent or limit inputs of pollutants into groundwater (Article 1.1). Regarding the assessment of the chemical status of groundwater, the Groundwater Directive required member states to establish the threshold values at the national level and at the level of the river basin district by December 22, 2008 (Article 3.5). The Directive also required member states to identify any significant and sustained upward trend in concentrations of pollutants or indicators of pollution found in bodies of groundwater identified as being at risk and define the starting point for reversing that trend (Article 5.1). When a pollution trend is realized, member states were required to reverse such trends which present a significant risk of harm to the quality of aquatic ecosystems or terrestrial ecosystems, to human health, or to actual or potential legitimate uses of the water environment (Article 5.2). In order to achieve the objective of preventing or limiting inputs of pollutants into groundwater, member states were required to establish program of measures (Article 6.1), and bring into force the necessary laws, regulations, and administrative provisions by January 16, 2009 (Article 12).

The requirements of this Directive were transposed into the Czech Republic by these legislative acts:

Act No.254/2001(Water Act) required the monitoring and assessment of the status of surface water and groundwater including the impact of human activities on such water and examination of the status of water bodies and ecological potential of heavily modified and artificial water bodies (Section 21.2.a). It gave this responsibility to the river basin administrators and other professional institutions, authorized, established, or founded for this purpose by the MoA or the MoE (Section 21.4). Regarding the periodical revisions of these assessments, the Act required a summary report on the fulfillment of programs of measures, the status of surface water and groundwater, and on water management in individual river basin districts to be submitted to the government every three years (Section 26.5). In case of nonachievement of the objectives for the status of surface water and groundwater, the Act stipulated that the causes of possible nonfulfillment must be examined (Section 26.3.a) and additional measures for achieving the aims of water protection must be adopted, including setting of stricter values for selected indicators or stipulating additional indicators, as appropriate (Section 26.3.d).

Government Decree No. 5/2011[39] also provided the transposition of the Groundwater Directive in the Czech legislation. The Decree stipulated that the groundwater status assessment shall consist of (a) the assessment of the status of groundwater bodies, (b) the assessment of significant and sustained upward trends and contamination clouds, and (c) the groundwater quality assessment (Article 5). Method of assessment of groundwater bodies shall include the assessment of the chemical and quantitative status (Article 6.1). For the assessment of the status of bodies of groundwater, the results obtained from the groundwater survey network, analyzes of general and water management characteristics of river basins, and the assessment of the impact of human activity on the status of groundwater bodies shall be used (Article 6.2). The groundwater quality assessment shall be carried out once a year for individual monitoring sites in accordance with the procedure defined in Part A of Annex 5 to this Decree (Article 10.1), and the reference values set out in Annex 5 to this Decree shall be used for quality evaluation (Article 10.2).

In accordance with the Framework Program for the Identification and Evaluation of the State of Waters and Protected Areas of the Czech Republic, the groundwater monitoring network was proposed and it was divided into groundwater chemical status monitoring network and groundwater quantitative status monitoring network (Article 15.1). The groundwater chemical status monitoring network was designed to provide a coherent overview of groundwater quality in a subbasin, to identify significant and sustained upward trends in pollutants and to assess the groundwater chemical status (Article 15.2). The groundwater quantitative monitoring network was designed to provide reliable data to evaluate the quantitative status of groundwater bodies or bodies of groundwater, including an evaluation of the usable quantities of groundwater, natural resources, groundwater regime, and groundwater yield (Article 15.3.a). Regarding the assessment of any upward trend, the decree required that the assessment of a significant and sustained upward trend is to be carried out for pollutants, groups of pollutants or values of pollution indicators in bodies of groundwater once every six years (Article 9.1).

The first cycle of the RBMP of the Czech Republic, approved by the Government Resolution No. 562, stated its policy objectives regarding the status of groundwater as to stop or reduce infiltration of pollutants into groundwater, prevent deterioration of the status of all groundwater bodies, monitor the development of the status and supplies of groundwater, and avert any significant and lasting trend of concentration of harmful substances as a consequence of human activity with the aim of reducing pollution of groundwater (Objective 1.1.2). As mentioned earlier, Article 6.1 of the EU Groundwater Directive required member states to establish a program of measures against lasting trends of water pollution. The choice of the means and measures to deliver compliance was left to the member states because the member states understand best the local water quality situation and they can provide the most appropriate answers to the local problems. The second cycle of the *RBMP* of the Czech Republic described the basic and supplementary measures as constructing or upgrading urban wastewater treatment plants (1,209 basic measures), reducing nutrient pollution from agriculture (4 supplementary measures), remediation of contaminated sites (275 supplementary measures), improving hydromorphological conditions of water bodies (438 supplementary measures), protection of drinking water (12 measures in total), and preventing or controlling the input of pollution from urban areas and transport (120 measures in total).[40]

This comparative analysis shows that the EU legislation on groundwater quality was transposed into the Czech Republic, and the requirement of the EU legislation was fulfilled by establishing threshold values for pollutants in groundwaters, carrying out groundwater quality assessment and assessment of significant and sustained upward trends for pollution, and devising programs of measures.

Council Directive 98/83/EC, Drinking Water Directive was prepared with an aim to protect human health from adverse effects of any contamination of water intended for human consumption by ensuring that it is wholesome and clean (Article 1.2). The Directive required member states to set values applicable to drinking water for the parameters set out in Annex I (Article 5.1) and to take all measures necessary to ensure that regular monitoring of the quality of drinking water is carried out in order to check that the water available to consumers meets the requirements of this Directive and in particular the parametric values set in accordance with Article 5 (Article 7.1). In order to meet these obligations, member states shall establish appropriate monitoring programs for all drinking water (Article 7.2).

This Directive was transposed in the Czech Republic by these legislative acts:

The Ministry of Health stipulated Act No. 258/2000[41] that regulated the hygenic requirements for water (Article 3); obligations of persons in control of drinking water and the conditions of drinking water supply (Article 4); the products, chemicals and chemical mixtures coming into contact with drinking water and the water treatment technology (Article 5); and the basic sanitary requirements for the natural and artificial swimming pools and saunas (Article 6). The Act obliged the Ministry of Health to draw up "the operating rules on the source and place of sampling of raw water; a monitoring program; and a risk assessment" (Article 3.c). The monitoring programme shall "provide information on the quality of drinking water; and verify that the measures work effectively throughout the water supply chain; and that the water is harmless to the customer and complies with all requirements set by the implementing legislation" (Article 3.c.2). The risk assessment shall include "description of the water supply system; description of the hazards identified and an estimate of their severity; and the establishment of corrective or control measures to eliminate or mitigate unacceptable risks throughout the supply system" (Article 3.c.5).

Government Decree No. 252/2004[42] laid down the sanitary requirements for drinking water and the frequency and extent of inspections of drinking water. Incorporating EU legislation, this Decree modified (a) hygienic limits of microbiological, biological, physical, chemical indicators of drinking water quality; (b) the extent and frequency of checking compliance with drinking water quality; (c) requirements for drinking water quality control methods (Article 1). Drinking water quality indicators were given in Annex. The control measures of the drinking water quality should include control of risk activities in the water resource protection zone or, where necessary, the wider associated area, in order to protect the water source from pollution (Article 4.1.a) and checks on the status of the protection zone, the construction and technical condition of the collection facilities, water treatment plants, water reservoirs, pumping stations, and other related infrastructure of the drinking water supply system (Article 4.1.b). The locations, scope, and frequency of inspections, measurements, and abstractions shall be determined by the complexity and vulnerability of the supply system (Article 4.2). The Decree stipulated that the operational analyzes, which include the collection and analysis of raw, treated, and supplied drinking water for the purpose of controlling the production of drinking water, shall be carried out according to the water quality control plan pursuant to Act on Water Supply and Sewerage Systems (Article 4.4).

This comparative analysis shows that the EU legislation on drinking water quality was transposed into the Czech Republic, and the requirement of the EU legislation was fulfilled by establishing sanitary requirements for drinking water and carrying out regular inspections for compliance with the established standards.

Conclusive Remarks

This legislative analysis demonstrated that the Czech Republic has transferred the EU legislation on water management into its domestic structure. Theoretical implication of this empirical analysis is that it validated the research hypothesis on policy transfer mechanisms. I put forward earlier that the soft transfer emerges when member states transfer the ideas, concepts, and strategies presented in the EU's policy documents, as opposed to the hard transfer that emerges when the EU transfers its rules and procedures to member states. While the first mechanism is related to the concepts of social learning or ideational learning, the second one is conceptualized as a mandatory requirement. The first part of this empirical research focused on the soft transfer through a content analysis of the executive portfolio of the EU and Czech Republic in water management policy, and concluded that the normative adaptation is evident. The second part of this empirical research focused on the hard transfer through a content analysis of the EU legislation and domestic legislation in a comparative perspective, and reached to the conclusion that legislative adaptation is also evident. These findings signify a domestic structural change in water management policy (see table 3.1).

Table 3.1 Comparison of the Policy Legislation in the EU and in the Czech Republic (legislative adaptation)

European Union's Legislative Portfolio	*Czech Republic's Legislative Portfolio*
Water Framework Directive (Directive 2000/60/EC)	Act No. 254/2001
Nitrates Directive (Directive 91/676/EEC)	Act No. 254/2001 Act No. 156/1998 Regulation No. 103/2003 Regulation No. 219/2007 Regulation No. 262/2012
Drinking Water Directive (Directive 98/83/EC)	Act No. 258/2000 Act No. 274/2003 Decree No. 252/2004
Urban Waste Water Directive (Directive 91/271/EEC)	Act No. 254/2001 Act No. 274/2001 Regulation No. 61/2003 Regulation No. 401/2015
Ground Water Directive (Directive 2006/118/EC)	Act No. 254/2001 Decree No. 5/2011

Source: Author's own compilation

IMPLEMENTATION OF THE POLICY NORMS AND LEGISLATION IN WATER MANAGEMENT

This part aims to analyze the implementation of the new policy norms and policy legislation in the Czech Republic. As discussed earlier, what this book offers as a contribution to the existing literature on Europeanization is to empirically test and validate the completeness of policy transfer through a statistical data analysis based on a set of indicators. Positive changes in these indicators will reveal successful operationalization of the normative and legislative adaptation, and therefore, fulfillment of the policy objectives and strategies in water management.

The set of indicators under consideration in this part is categorized into four areas:

1. Status of surface water
2. Status of groundwater
3. Public water supply
4. Wastewater

Regrettably, some other indicators had to be discarded from the data set for the sake of clarity and efficiency of this research. When using the statistical data, a mix of ex-ante (factual) and ex-post (using only available data) approach is adopted. The statistical data used here comes from the national and European statistics. Only the official statistics that are provided by the relevant governmental agencies of the Czech Republic and by Eurostat are used to analyze and compare the detectable and empirically provable domestic changes in this policy area. The main principle here is that the data chosen for comparison is obtained from periodic surveys or other sources regularly providing reliable data in the time series.

Status of Surface Water

Article 8(1) of the WFD requires member states to establish monitoring programs for the assessment of the status of surface water in order to provide a coherent and comprehensive overview of water status within each River Basin District. These programs shall cover the ecological and chemical status for surface waters, and the chemical and quantitative status for groundwaters.[43]

Ecological status is an assessment of the quality of the structure and functioning of surface water ecosystems. It shows the influence of pressures such as pollution and climate change on a range of quality elements. These are primarily biological quality elements with supporting hydromorphological and chemical and physico-chemical quality elements.

In 2018, European Environment Agency Report on the assessment of status and pressures on European waters concluded that in EU-28 (except Greece, Ireland, and Lithuania) around 40 percent of the surface water bodies are in high or good ecological status, while 60 percent did not achieve good status.[44] The Scandinavian region and Scotland, as well as Estonia, Romania, Slovakia, and several RBDs in the Mediterranean region achieved to maintain a high proportion of their water bodies in high or good ecological status. In contrast, many of the central European RBDs, as well as Hungary, have the highest proportion of water bodies that are not in good ecological status.[45]

As a comparison to these figures, the Czech Republic reported in its second RBMP that 19.2 percent of its surface water bodies were at good (18.8%) or high ecological status (0.4%).[46] Around 21 percent of surface water bodies were expected to be at least good ecological status by 2015, but 100 percent achievement was not expected until 2027.[47] When we seek for the trend in the ecological status of the Czech surface waters from the first RBMP to the second one, we find that overall ecological status has improved since the first RBMP with considerably fewer water bodies in poor/bad status and more water bodies in moderate status. In a regional perspective, in the Danube there is an increase in the number of good status water bodies, but only a minor increase is seen in the Elbe and a decrease in good status water bodies in the Oder.[48]

The main pressures on surface water bodies are hydromorphological pressures, diffuse source pollution, point-source pollution, and water abstraction. According to 2018 data, in the EU-28 (except Greece, Ireland, and Lithuania) diffuse sources affect 38 percent of water bodies, and the main diffuse source pressure is agriculture, which affect 25 percent of all surface water bodies. Point sources affect 18 percent of water bodies, and the main point-source pressure is discharges from urban wastewater treatment plants, which affect 12 percent of all surface water bodies. Hydromorphological pressures affect 40 percent of water bodies, and the main pressures are dams, which affect 24 percent of all surface water bodies, and physical alteration of channel, which affect 26 percent of all surface water bodies. The impacts of morphological changes on surface water bodies are nutrient enrichment and altered habitats. It is important to note that a water body may be affected by more than one pressure.[49]

As a comparison to these figures, the Czech Republic reported in the second RBMP that the most significant pressure on surface of water bodies was anthropogenic pressure, which affected 70 percent of surface water bodies. The main point-source pressure was urban wastewater, which affected 38 percent of surface water bodies. The main sources of diffuse source pollution were agriculture (22%), atmospheric deposition (22%), and discharges not connected to the sewerage network (19%). The main hydromorphological pressure was the physical alteration of channel/bed, which affected 26 percent of all surface water bodies. When we seek for the trend for these pressures between the two cycles, we find that the same broad types of pressures seem to have been reported for both cycles.[50]

Surface water quality is affected primarily by point and diffuse sources of pollution caused by discharges from wastewater treatment plants, industrial effluents, and agricultural runoff. These sources of pollution may have devastating effects on water bodies. Many organic materials, including sewage effluent and farm and food-processing wastes, consume oxygen, and suffocate fish and other aquatic life. Excess nutrients can create eutrophication, a process characterized by increased plant growth, problematic algal blooms, depletion of oxygen, loss of life in bottom water, and undesirable disturbance of the balance of organisms present in the water. Pollution through hazardous substances and chemicals can threaten aquatic ecosystems and human health.

The organic matter indicators of water pollution are BOD_5 (biochemical oxygen demand) and COD_{cr} (chemical oxygen demand) that are used to assess the amount of organic oxygen-consuming pollution in a river. There are also nutrients, such as nitrogen and phosphorus. Nitrogen occurs in surface waters in organic and inorganic forms, namely nitrate (NO_3-N), ammonia (NH_4-N), and nitrite (NO_2-N). In order to improve the quality of surface water bodies and maintain a sustainable water environment, it is important to reduce pollution measured by these indicators. In this section, we will comparatively analyze annual statistics to find out how successfully the EU-zone and the Czech Republic managed to achieve this goal.

Statistical data shows that in European rivers, BOD_5 concentrations decreased between 1992 and 2017 with an average annual decrease of 0.07 mg/l (-2.5% per year). This trend has been stabilizing for the past five years. Likewise, there was an average decrease in annual ammonium concentration of 9.1 µg/l (-3.5% per year). The shorter time series include more monitoring sites, so it is better representative for the European rivers (see figure 3.1).

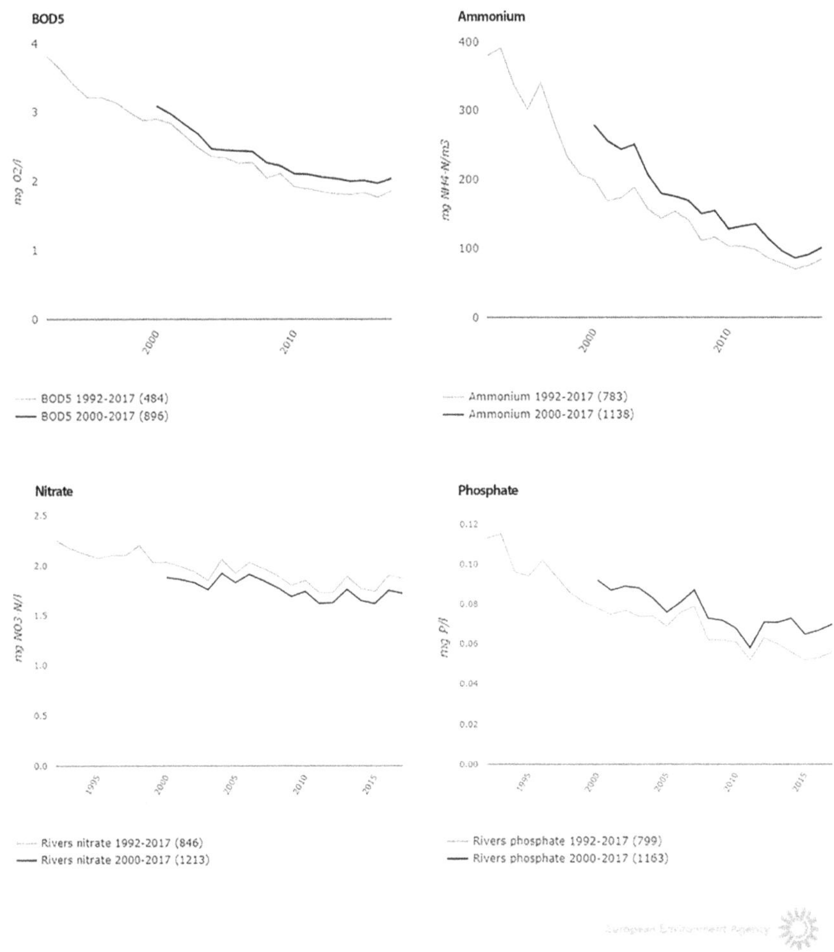

Figure 3.1 Trends in Biochemical Oxygen Demand, Ammonium, Nitrate, and Phosphate in European Rivers. *Source*: European Environment Agency, Indicator Assessment.

This positive achievement can be attributed to improved biological treatment of wastewater and implementation of the UWWTD and other relevant directives.[51]

The average level of nitrate in European rivers decreased steadily over the period 1992–2009 but has varied around the same level since then. Agriculture is the largest contributor of this pollution. The EU Nitrates Directive and other related EU directives[52] helped to reduce the nitrogen pollution from agriculture over the last decades. However, in order to deal with the stabilization of river nitrate concentrations, the EU may require further measures to be taken by the member states.

The average phosphate concentration in European rivers has decreased markedly over the last decades (by 1.6% per year). The decrease in the level of phosphate concentrations in rivers can be attributed to the measures introduced by the European legislation, in particular the UWWTD[53] and its successful implementation by the member states. Also, the change to phosphate-free detergents has contributed to lower phosphorus concentrations.[54]

As a comparison, we will analyze the trends for the same pollution indicators in the watercourses of the Czech Republic. Statistical data shows that, during the period 2000–2018, the BOD_5 indicator decreased by 18.0 percent and COD_{Cr} decreased by 3.2 percent. In the same period, the Czech Republic managed to successfully reduce $N\text{-}NH_4$ levels by 70.2 percent and P_{total} levels by 34.4 percent. A significant deviation is seen in 2003 which was caused by the extreme flood situation; since then, the decreasing trend has continued. The value of $N\text{-}NO_3$ decreased by 15.9 percent in comparison with 2000. This empirical analysis revealed that the Czech Republic succeeded reducing the quantity of the nutrients (nitrogen and phosphorus) in its watercourses. The reduction of this organic pollution can be attributed to the construction and modernization of municipal and industrial wastewater treatment plants that apply biological nitrogen removal and biological or chemical phosphorus removal and also to the reduction of phosphates in detergents[55] (see figure 3.2).

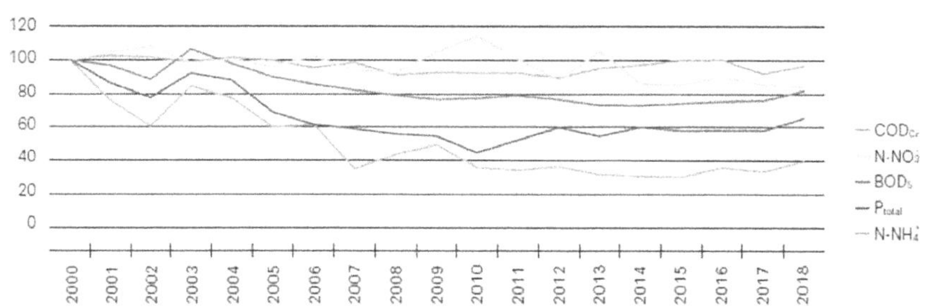

Figure 3.2 Development of Concentrations of Pollution Indicators in Czech Watercourses (2000–2018). *Source*: Ministry of Environment, Report on the Environment of the Czech Republic 2018, p. 76.

For surface waters, *chemical status* is defined based on the concentrations of certain pollutants. Directive 2008/105/EC[56] defined thirty-three priority substances or groups of substances. Good chemical status means that no concentrations of priority substances exceed the relevant standards established in this Directive. Member states are required to monitor all priority substances which are discharged into the river basin or subbasin, and undertake sufficient amount of monitoring in order to obtain a reliable and robust assessment of the chemical status of all water bodies in the RBD.

Reporting under the second RBMP shows that in the EU-28 (except Greece, Ireland, and Lithuania), 38 percent of surface water bodies were in good chemical status (by number of water bodies), while 46 percent were not achieving good status and the status of 16 percent was unknown.[57] There are also substantial differences between member states. The results of monitoring of all priority substances revealed that in some member states, over 90 percent of surface water bodies failed to achieve good chemical status (Austria, Belgium, Germany, Slovenia, Sweden), while in others (Croatia, Cyprus, Romania, Slovakia, Spain, the United Kingdom) 10 percent or less of surface water bodies failed to achieve good chemical status (leaving aside the countries with high percentage of unknown water bodies). These substantial differences between member states may result from the differences in population density, industry, geography as well as the approach taken in monitoring.[58]

As a comparison to these figures, in the Czech Republic, the second RBMP reported that of 1,121 surface water bodies, 69 percent achieved good chemical status, while 31 percent failed to achieve good status based on the most recent assessment in 2013. For the assessment of chemical status, thirty substances were monitored at 80 percent of the sites, and cadmium at 6 percent of sites, with a monitoring frequency ranging from six to twelve samples per year.[59]

Among the EU member states, the most frequently reported priority substances found in surface water bodies were mercury, brominated diphenyl ethers, benzo(g,h,i)perylene, benzo(a)pyrene, fluoranthene, cadmium, and nickel.[60] As a comparison, in the Czech Republic, the largest proportion of exceedances were found for fluoranthene (in 23% of surface water bodies), benzo(g,h,i)perylene (in 20% of surface water bodies), benzo(a)pyrene (in 18% of surface water bodies), nickel (16%), and mercury (7%).[61] This comparative analysis helps to understand the relevance of priority substances among member states.

Status of Groundwater

The WFD requires *good quantitative status* to be achieved by ensuring that the long-term annual average rate of abstraction does not exceed available groundwater resource. The main pressure causing failure to achieve good quantitative status is water abstraction for public water supply, agriculture, and industry.

In the second RBMPs, in the EU-28 (except Greece, Ireland, and Lithuania), around 90 percent of the area of groundwater bodies was reported to be in good quantitative status, 9 percent had poor quantitative status, while around 1 percent of the groundwater body area had unknown status. Six member states (Austria, Latvia, Luxembourg, the Netherlands, Romania, and Slovenia) reported that all of their groundwater bodies were in good quantitative status, while Cyprus and Malta reported the highest proportion of groundwater bodies in poor status, at 57 percent and 80 percent, respectively. Both of these countries depend heavily on groundwater resources to meet their water needs, with Malta abstracting around 60 percent from this source and Cyprus abstracting almost half. Fourteen member states (Croatia, Denmark, Estonia, Bulgaria, Portugal, Germany, Poland, Finland, Sweden, Czech Republic, France, the United Kingdom, Spain, and Italy) reported that between 75 percent and 99 percent of the total area of groundwater bodies were in good quantitative status, while three gave a figure of between 50 percent and 75 percent (Hungary, Slovakia, and Belgium).[62]

As a comparison to these figures, in the Czech Republic, the assessment of quantitative status of groundwater bodies showed that 69 percent of groundwater bodies were in good quantitative status, 10 percent were failing good status, and 21 percent had unknown status.[63] In 2018, water supply for public use accounted for 80.8 percent of groundwater abstraction. The industry accounted for 9.9 percent of groundwater abstractions.[64]

In accordance with the WFD, good *groundwater chemical status* is achieved when (1) there is no sign of saline intrusion in the groundwater body; (2) the concentrations of pollutants do not exceed the permitted or threshold values under the groundwater quality standards; and (3) there is no failure in achieving ecological or chemical status of associated surface waters, or any significant damage to terrestrial ecosystems that depend directly on the groundwater body. Member states are required to carry out an initial characterization of all groundwater bodies to assess their uses and the degree to which they are at risk of failing to meet the objectives for each groundwater body under Article 4. The chemical status of the groundwater bodies is assessed as good or failing to achieve good status according to its compliance with EU standards for nitrates (50 mg/l) and pesticides (0.1 μg/l for individual pesticides, total maximum 0.5 μg/l), and with member states' established threshold values for other groundwater pollutants.

EEA's Assessment of European Waters in 2018 revealed that in the EU-28 (except Greece, Ireland, Lithuania, and Slovenia), 74 percent of groundwater bodies (by area) were in good chemical status and 25 percent had poor chemical status, with 1 percent of unknown status. Lowest groundwater quality was found in areas where there is intensive agricultural production and heavy industry. Through pollution from nitrates and pesticides, agriculture became the main pressure causing failure to achieve good chemical status in groundwater in the EU-zone. Nitrates were reported by twenty-four member states as the main pollutant that most commonly cause poor chemical status (causing failure in 18% of groundwater body area); and pesticides were reported by eleven member states as the second major pollutant (causing failure in 6.5% of groundwater bodies by area).[65]

As a comparison with these figures, in the Czech Republic, national groundwater quality monitoring network analyzed 691 monitoring sites in 2018 (the number of monitoring sites increased from 653 in 2010). The assessment is carried out in accordance with Decree No.5/2011 which established limit values of indicators for the groundwater quality assessment. When we analyze the trend in chemical status of groundwater in the time period 2010–2018, we find that in 2010, of 653 monitoring sites assessed, 85.6 percent of sites exceeded limit values for at least one indicator, while in 2018, 83.5 percent of 691 monitoring sites exceeded limit values for at least one indicator.[66] This chemical status is unsatisfactory to a large extent.

According to the 2018 Report of the Ministry of Environment of the Czech Republic, main pollutants of groundwater are ammonium ions (11.0% of samples were above the limit) and nitrates (10.6% of samples were above the limit). Excess concentrations of various pesticides are reflected in 22.6 percent of samples.[67] The main reason for contamination of groundwater with pesticides is the trend of growing energy crops such as rape and maize that are treated with herbicides. Regarding the trend in the last three years for the groundwater quality, the status in 2018 did not change significantly compared to the previous three years, 2015–2017, due to the slow dynamics of changes in groundwater chemistry.

Public Water Supply

In 2018, in the Czech Republic, water supply systems supplied water to 10.064 million inhabitants, that is, 94.7 percent of the total population. The corresponding data in the year 2003, one year before accession to the EU, was 89.8 percent of the population. The highest percentage of population supplied with water from water supply systems in 2018 was recorded in the City of Prague and in Karlovy Vary region with 100 percent of the population, followed by the Moravian-Silesian region with 99.9 percent of the population. The lowest percentage of inhabitants supplied with water was recorded in the Pilsen region (84.4%) and in the Central Bohemian Region (86.4%).[68]

When we compare this data with those from other EU member states, we find similar results. According to Eurostat statistics, among the countries that reported their national data, the highest percentage of population supplied with water from water supply systems was recorded in Spain, the Netherlands, Hungary, Malta, and Cyprus (100% in 2017), which were followed by Germany (99.3% in 2016), Poland and Portugal (92% in 2017), and Slovakia (88.9% in 2017).[69]

In the past, European water management policies have largely focused on increasing water supply by, for example, drilling new wells, constructing dams and reservoirs, and building large-scale water-transfer infrastructures. However, it is not possible to achieve an endless increase of the water supply, so there is a new focus on limiting the pressures on it. The main pressures on water supply come mainly from water abstractions (for public water, agriculture, and industry), and these pressures are particularly significant in those areas that are prone to low precipitation levels and drought. Annual statistics provided by Eurostat show that the highest amount of freshwater abstraction by public water supply in 2017 came from Greece (179.2 m^3 of water per inhabitant), which was followed by Ireland (127.5 m^3 per inhabitant) and Bulgaria (124.2 m^3 per inhabitant). These patterns reflect specific conditions in these countries: in Ireland the use of water from the public supply is still free of charge for many households, while in Bulgaria there are high losses from the public network. On the other hand, the lowest figures were provided by Malta (31.2 m^3 per inhabitant), Estonia (44.8 m^3 per inhabitant), and Lithuania (45.3 m^3 per inhabitant). The Czech Republic also scored low (with 58.3 m^3 use of water per inhabitant) compared to other EU member states.[70] These various results may stem from one or several factors combined, such as the reduction of water losses through improved maintenance of networks, use of water-saving household appliances, and increased level of awareness concerning the cost or value of water and the environmental consequences of wasting it.

In the Czech Republic, water loss from the water supply network is still a problem although it has decreased significantly in the period 2010–2018. The share of drinking water losses from the water supply network was 25.2 percent in 2010, but it decreased to 15.8 percent in 2018.[71] These losses are caused by accidents and leaks from water mains due to old infrastructure.

EU member states are obliged under the Drinking Water Directive to monitor on a regular basis the quality of the water that is supplied by public water supply systems to consumers. The Drinking Water Directive distinguishes three groups of parameters as microbiological parameters, chemical parameters, and indicator parameters.[72] The Czech Republic achieved high compliance rates[73] in all parameter groups reaching a rate of 99.93 percent for microbiological parameters, 99.93 percent for chemical parameters, and

99.32 percent for indicator parameters in 2013.[74] When we compare this data with the EU-average, we find that the results of assessments[75] that were carried out in the EU member states within 2011–2013 period revealed high compliance rates of over 99 percent for microbiological and chemical parameters.[76] The achievement of high compliance rates both in the EU member states and in the Czech Republic reflects successful implementation of the Drinking Water Directive and the fulfillment of its policy goals.

Wastewater

The treatment of urban wastewater from our homes and workplaces is fundamental to ensuring public health and environmental quality. Article 3 of the Urban Waste Water Treatment Directive (Directive 91/271/EEC) requires member states to ensure that all agglomerations above 2,000 population equivalent are equipped with sewerage systems for urban wastewater. The Directive lays down criteria for specific types of treatment: Article 4 requires that municipal wastewaters drained by sewerage systems shall undergo secondary treatment or other equivalent treatment before discharge. Article 5 requires that in catchments with particularly sensitive waters, such as those suffering from eutrophication, more stringent tertiary wastewater treatment may be required to substantially reduce nitrogen and phosphorus pollution.

When we analyze the compliance of the Czech Republic to these standards, we find that the EU accession had a major impact on the development of infrastructure for the collection and treatment of sewerage and for the subsequent compliance with the European legislation. When the time period 2000–2018 is analyzed, we find that in 2003, the last year before joining the EU, the proportion of the Czech population connected to the sewerage networks (with or without a WWTP) was 77.7 percent, and this rate increased to 85.5 percent in 2018. There is also a positive development in the proportion of population that is connected to the sewerage systems with WWTP from 68.9 percent in 2003 to 82.4 percent in 2018 (see figure 3.3).[77]

The Czech Republic has also achieved more than a twofold increase in the number of WWTPs in the time period 2002–2018 (see figure 3.4). The statistical data reveals that in 2002 there were 40 primary treatment facilities, 694 secondary treatment facilities, and 500 tertiary treatment facilities in the Czech Republic, whereas in 2018, these numbers reached to 25 primary treatment facilities, 1,155 secondary treatment facilities, and 1,497 tertiary treatment facilities.[78] This development shows that the impact of the EU accession on wastewater treatment policy is discernible, and the normative and legislative changes in this policy area have been operationalized in the practice.[79]

As a comparison, when we look at the same figures in other EU member states, we find that there is a development toward a higher proportion of the

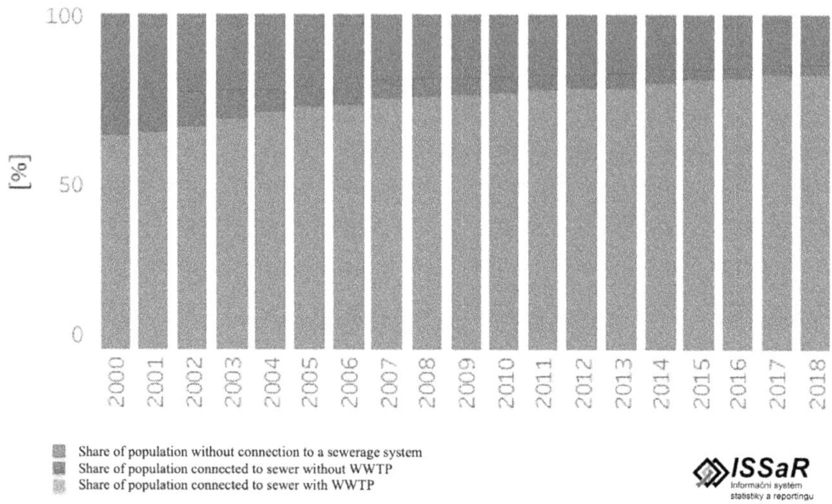

Figure 3.3 Percentage of Inhabitants Connected to Sewerage System in Czech Republic (2000–2018). *Source*: CENIA.

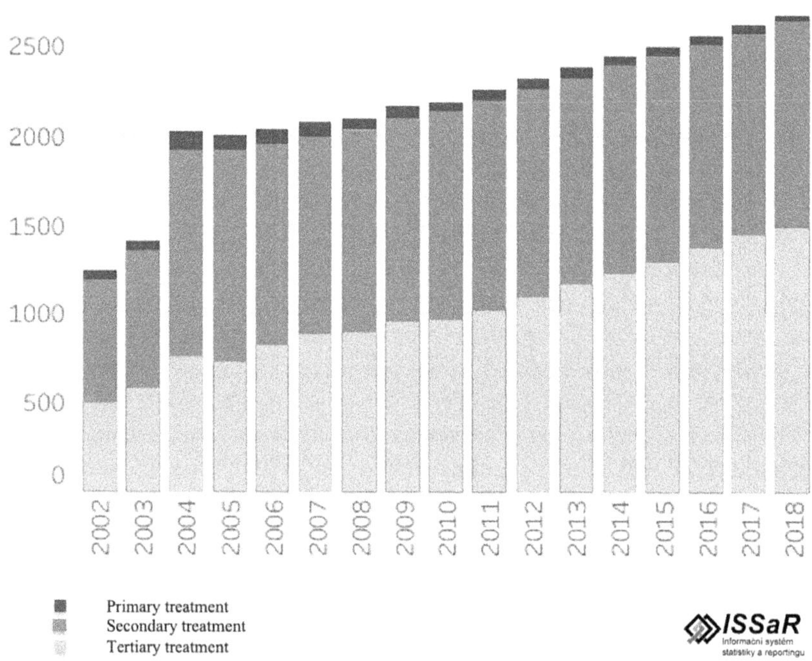

Figure 3.4 Number of Wastewater Treatment Plants According to Treatment Stages in the Czech Republic (2002–2018). *Source*: CENIA.

population being connected to wastewater treatment of any level. Malta is the frontrunner in such development with the coverage reached 100 percent in 2011—up from around 20 percent in 2010—due to the construction of new wastewater treatment plants. Naturally, the growth of the coverage rate is slower in the larger countries, but the situation is increasing everywhere. The recent Eurostat data shows that above 80 percent of the population in sixteen of the EU member states[80] (mixed reference years 2016 and 2017) is connected to wastewater treatment of any level in compliance with Article 3 of UWWT Directive. The share of population connected to at least secondary wastewater treatment plant is above 95 percent in seven member states (Germany, Latvia, Luxembourg, the Netherlands, Austria, Sweden, and the United Kingdom) (mixed reference years 2016 and 2017) in compliance with Article 4 of UWWT Directive.

The statistics show that, in the Czech Republic, total volume of discharged wastewater has exhibited a downward trend since the 1980s, with only occasional occurrence of a year-to-year increase in volume. In 2018, in total 1.5 million m³ of wastewaters and mine waters were discharged into surface waters (see figure 3.5). Compared to 2016 and 2017, this represented a decrease of 10 percent.[81] The structure of the wastewater discharges in the Czech Republic shows that the largest share of discharges is from the public sewerage systems and energy. The largest percentual decrease in the quantity of discharged wastewaters was observed from 1997 to 2001 in the energy sector, although it started to increase again after this date. This sudden increase after this date is related to the water used for the cooling of the nuclear power plants Temelin and Melnik. Another continuous decrease was observed in the

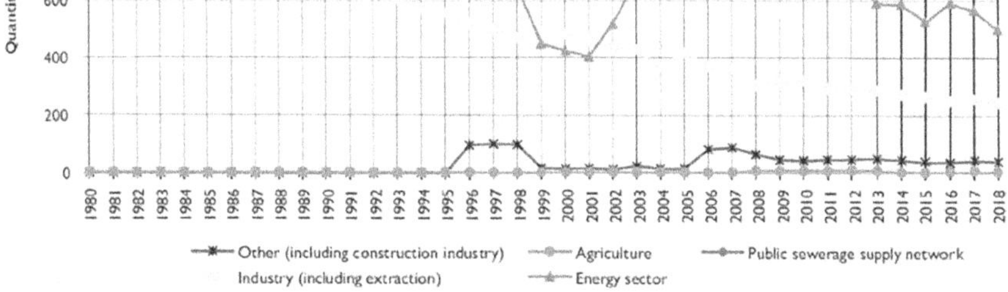

Figure 3.5 Discharge of Wastewaters in the Czech Republic (1980–2018). *Source*: Ministry of Agriculture and Ministry of Environment, Report on Water Management in the Czech Republic 2018, p. 33.

category of public sewerage systems. The exceptions were the years 2010 and 2013, when there was a significant increase in discharges (sewerage and rain water). This was caused by higher precipitation which increased the volume of discharged rain water. After these years, the volume of discharged wastewater continued to decrease. While wastewater discharges in energy sector influence water temperature and oxygen regime, municipal wastewater discharges are important point sources of pollution, mainly organic. Statistical data shows that agriculture discharged only minor volume of wastewater in the period under consideration. The construction industry, except for a temporary increase from 1995 to 1998 and from 2005 to 2008, recorded the second lowest discharge of wastewater.[82]

Best Practices for Water Management

In the United Kingdom, the most significant water quality problem is diffuse pollution. In rural areas, agriculture and rural land management collectively cause pollution to water bodies. In urban areas, a range of substances from sewage overflow pollute the urban water environment. In order to reduce pollution and maintain a satisfactory water quality, water companies began to explore different approaches with land managers in the early 2000s. As a result, two large projects were developed: Sustainable Catchment Management Programme or SCaMP (introduced in 2005 by United Utilities in the North West of England), and the Upstream Thinking project (introduced in 2008 by South West Water).[83]

Upstream Thinking is a sustainable approach bringing together the expertise of partners and the knowledge of farmers to improve water quality at source. Farm advisers visit farms and carry out an assessment resulting in a whole-farm plan. This includes a water management plan and includes future capital investment proposals targeted at water quality improvements, which will be up to 50 percent funded by Upstream Thinking. These can include improvements to slurry storage, fencing to keep livestock out of rivers, providing alternative water sources for livestock, and better pesticide management including investment in new equipment such as weed wipers. Monitoring of the program showed increases in plant and animal life like dragonfly and mayfly which are indicators of good water quality.

The aim of SCaMP project is to benefit both water and wildlife through improved catchment management. SCaMP 1 (2005 to 2010) included projects across 27,000 hectares of water catchment areas. In this context, rehabilitation of moorlands to protect sensitive habitats, improving water retention, and thereby, reducing erosion and sediment losses were key actions. Following the success of the first project, SCaMP 2 (2010 to 2015) was introduced across 30,000 hectares. After several years of water quality monitoring, SCaMP project was shown to have significant beneficial effects on water quality in key locations.

With an aim to reduce water pollution from pesticides, France launched the Ecophyto plan[84] in 2008. The aim of the plan was to reduce the use of pesticides by 50 percent by 2018 and to encourage farmers to use alternative ways of controlling crops. Dissemination of information on low-pesticide methods and improvement of current techniques were the first actions to be implemented. Demonstrations of new low-pesticide methods were provided through the mobilization of all partners in research, development, and knowledge transfer. The partnerships between research, development, and industry professionals were reinforced in order to create innovative design and development of low-pesticide technical pathways and cropping systems. Continuous training programs were developed not only for users but also for distributors, advisers, and operators in order to encourage reduced and safe pesticide use. In this context, introduction of revised vocational qualifications for initial entry into farming from the autumn of 2008 was planned. Also, a new form of governance was consolidated through involvement of public- and private-sector actors alongside nonprofit associations and regional actors.

In the framework of the plan, technical assistance was provided to 3,000 volunteer pilot farms enrolled as peer groups since 2011. In practice, engineers conducted an initial diagnosis of farmers' practices and then worked with them to draw up a plan to reduce their use of pesticide over five years. They supported the farmers in implementing the project, and monitored its progress through campaign reviews and annual documentation. After four years in the program, participating farms used 8 percent to 22 percent lesser pesticides than they would have used in the absence of the program. The reduction in the use of chemicals was accompanied by an increase in the use of biocontrol products ranging from 24 percent to 33 percent. This change of practices resulted in a reduction in yields for a fraction of enrolled farms while others maintained their yields.[85]

Version II+ of Ecophyto plan, updating and improving the previous plan, was introduced in 2018. The new plan maintained the goal of reducing the use of plant protection products by 50 percent by 2025.

In Copenhagen, Denmark, the new Ørestad district represents an innovative approach to water management. Here, the distinct use of water for aesthetic purposes is directly linked to the storm water management. With a three-pronged system, household wastewater is discharged into a central treatment plant, roof water is discharged into recreational canals, and road water is treated locally before being discharged into the country's recreational canals. A new treatment technology called Dual Porosity Filtration is developed for the cleaning of road runoff. Thus, storm water runoff becomes the main supply for the 10 km of canals that form the unique blue structure of Ørestad.[86]

A similar approach was taken in Stockholm, Sweden, by a district called Hammarby Sjöstad. The change started with the definition of new environmental goals for sustainable water management. The first goal was to reduce residential water consumption by 50 percent: Stockholm average was 200 liters/person/day, and Hammarby's goal was 100 liters/person/day. In order to achieve this goal, proper education for the district's residents and the use of water-saving appliances were used. The residents were also given the option of monitoring their water consumption via internet to increase their awareness of their habits. The second goal was to achieve cleaner wastewater: reduce the level of hazardous substances (heavy metals and contaminants) in the wastewater stream by 50 percent of other local systems, and capture 95 percent of the phosphorous content of the wastewater stream for use in agriculture. To achieve these goals, Hammarby employed several innovative techniques for large-scale water management and treatment: Catchment and settling basins were strategically located to capture any runoff from streets, buildings, walkways, and even the adjacent ski resort. Two massive, enclosed holding basins were capable of storing all locally generated storm runoff for a short period to allow for the settling of sediments and pollutants. An experimental wastewater treatment plant, Sjöstadsverket, was built as a working laboratory for new and innovative techniques for biological, chemical, and biogas producing waste treatment methods. A massive network of tiered drainage and settling canals were built throughout the district, providing the aesthetic beauty of natural creeks and vegetated swales.[87] From the 1990s to today, the district was revitalized from a polluted and unsafe industrial and residential area into one of Stockholm's most pleasant residential districts.

CONCLUSIVE REMARKS

Empirical Conclusions

In this chapter, we have seen significant adaptation of the Czech Republic's normative and legal structure to that of the EU in the water management area. Regarding *normative adaptation*, the comparative analysis demonstrated that the EU policy norms and policy strategies have been significantly transposed into the Czech Republic's national strategies and action programs.

Regarding *legislative adaptation*, the Czech Republic undertook to incorporate the requirements stipulated in the EU regulations into the national legislation. The 2001 Water Act, the main law for protection and management of water resources, included provisions for water conservation, economic instruments for water management, water planning, and international cooperation.

The Czech Republic also prepared RBMPs for its river basins to comply with the WFD, and used the national Water Act as the main implementation tool. The main legislative measures aimed at reducing diffuse water pollution from agricultural sources include Government Order No. 262/2012 on delimitation of vulnerable areas and a related program seeking to reduce and prevent nitrate pollution in such areas. These measures ban nitrogenous fertilizers and provide for crop rotation, soil erosion monitoring, and manure storage facilities. National Development Plan and Regional Development Plans for water supply and sewerage systems represent a medium-term planning of strategies and measures regarding water supply and sewerage systems. As a conclusion, the Czech Republic has well-developed and fairly complete policy and legal frameworks for water management, supported with quantitative targets.

This study underlines the significance of the implementation of policy objectives and policy legislation that is acquired as a result of policy transfer. We argue that the extent of the operationalization of policy objectives, strategies, and legislation constitutes the final indicator of the domestic change. Based on this approach, an empirical analysis of the implementation phase is carried out based on national and international statistics. This analysis provides the completeness of policy transfer. Indeed, the primary objective of the EU water management policy, that is achieving and maintaining clean and unpolluted water, can only be achieved by the member states to correctly implement and enforce EU directives and regulations.

This study analyzed several indicators of assessment of water quality in the Czech Republic in a comparative perspective with other EU member states. Water pollution is caused by several factors: many organic materials, including sewage effluent and farm and food-processing wastes, consume oxygen, and suffocate fish and other aquatic life. Excess nutrients can create eutrophication, a process characterized by increased plant growth, problematic algal blooms, depletion of oxygen, loss of life in bottom water, and undesirable disturbance of the balance of organisms present in the water. Pollution through hazardous substances and chemicals can threaten aquatic ecosystems and human health.

The water pollution was among the most pressing issues of the Czech environment at the beginning of the 1990s. The major waterways were heavily polluted. Attention was focused on reducing the amount of pollution discharged into surface water via construction and rebuilding, and intensifying the wastewater treatment plants as well as construction of sewer systems. As a result, pollution of waterways dropped substantially in the 2000s. When the situation in the early 1990s is compared with the most recent one in 2018, it is noticed that the status of watercourses in most sections has obviously improved. In 2018, most watercourses belonged to Classes I and II (unpolluted or slightly polluted water) and III (polluted water) based on indicators BOD_5, COD_{Cr}, $N\text{-}NO_3$, $N\text{-}NH_4$, P_{total}. However, there are still sections (although short ones) of watercourses that belong to Class IV (heavily polluted water) and Class V (very heavily polluted water).[88] In a regional perspective, in the Danube there is an increase in the number of good status water bodies, but only a minor increase is seen in the Elbe and a decrease in good status water bodies in the Oder. The analysis of the second RBMP shows that there has been a progress in the status of water quality elements and pollutants from the first RBMP. In particular, the ecological status of surface waters has improved.

According to the latest assessment of the European Commission, chemical pollution is the most significant impact on the surface water (48% of surface water bodies) and on the groundwater (73% of groundwater bodies) in the Czech Republic. This is followed by nutrient pollution (41% of surface water bodies and 53% of ground water bodies).[89] Agriculture contributes to diffuse pollution in 22 percent of the surface water and 61 percent of the groundwater bodies in the form of runoff from farmland.[90] Nitrate levels at many monitoring points also remain an issue, as does eutrophication. This shows that Czech Republic needs further measures to achieve the objectives set in the WFD. Particularly, increased attention should be given to the application of action programs in those areas where quality of the waters in the context of the nitrate directive is insufficiently improving.

A significantly positive assessment is that the drinking water is of very high quality with the microbiological and chemical conformity rates reaching 99.9 percent in compliance with the Drinking Water Directive (Directive 98/83/EC). Finally, the Czech Republic is in compliance with the Urban Waste Water Directive (Directive 91/271/EEC) with 85.5 percent of the population connected to the urban wastewater collecting system in 2017 and 2018, and 82.6 percent of the population connected to wastewater treatment plants with secondary treatment in 2017.[91]

As a conclusion, better implementation of programs of measures can be achieved in the Czech Republic by establishing good coordination between public administration and other stakeholders and by improving the monitoring of the effectiveness of programs.

A Czech policy-maker[92] in water management sector was interviewed on the implementation of the EU water management policy in the Czech Republic. In our interview, he raised the issue of achievement of good ecological potential of water sources as "one of the most difficult requirements of the WFD because such changes take a long time." Indeed, fulfillment of these policy objectives shows a very slow progress in all EU member states. Thus, he finds the achievement of the goals of the WFD in 2027 "not a reliable objective." However, he argues that "the Commission is unable to accept the change or improvement of the policy content with this consideration."

We also talked about several EU policy objectives and strategies that the Czech policy-makers in water management area find very useful. One of them is the protection of the quality of the surface water and groundwater and drinking water. Considering the EU legislation in this area, he pointed out that "EU policy strategies improved the water quality in European rivers and lakes, and the obligation for construction of WWTPs has also been an extremely positive step. Therefore, successful implementations of these policy strategies have been realized."

A final issue we discussed was the recently issued Commission proposal for a regulation on minimum requirements for water reuse.[93] The proposal refers to the challenge of providing availability of water in sufficient quantity and quality in line with the growing needs of the populations and the impact of climate change in Europe. Since water over-abstraction, in particular for agricultural irrigation, is one of the main threats to the EU water environment, as a response to this problem, the Commission suggested treated wastewater from urban WWTPs to be reused for agricultural irrigation purpose. The Commission finds this approach in full compliance with circular economy objectives, and puts forward that in the context of adaptation to climate change, water reuse for agricultural irrigation has the highest potential for the alleviation of water scarcity in Europe.

However, the Czech policy-maker finds this decision "controversial to the general effort of the EU legislation to prevent the contamination of the environment by micropollutants." Since the standard WWTP technologies do not remove the micropollutants (such as antibiotics, various pharmaceutical products, especially hormonal nature, cosmetic products, etc.), he believes, "the use of purified water for irrigation will definitely load the agriculture ground by these dangerous substances. As there is little information on the residues of these micropollutants in the soil and their passage to the plants, this method may harm human health." He accepts that the circular economy includes the water reuse; however

> the re-use of the treated water can be a plausible strategy only when a better water quality is achieved through improvement of the existing wastewater treatment technology. As long as there is no such technological development, some EU member states, including the Czech Republic, will apply derogation for the implementation of the water re-use.

Theoretical Implications

These empirical findings points out significant structural change in water management policy of the Czech Republic. This section briefly summarizes the theoretical implications of these empirical findings for each of the research hypotheses.

In the bureaucratic organization, we witnessed establishment of new institutions dealing with environment and water management as a result of post-Communist transition, but not as a result of EU accession. After the EU accession, existing institutional structure continued in the water management because it was perceived as adequate to handle the adoption of the EU's acquis. Thus, we don't observe bureaucratic reorganization in the water management area as a result of EU accession. However, there has been an obvious change in the administrative procedures in the environmental policy in general, and in the water management policy in particular. This change has taken place through the adoption and implementation of EU's methods for assessment, information, planning, auditing, and sanctioning as discussed in the previous section. This section demonstrated first, a discernible normative adaptation of the Czech Republic's water policy norms, policy objectives, and strategies to those of the EU, and secondly, a complete transposition of the contents of the EU directives in water management. Regarding the operationalization of such systemic changes, this chapter also showed that these newly adopted policy norms, objectives, strategies, and legislation have been implemented extensively in the Czech Republic, although with some shortcomings and gaps. All these empirical findings validate a structural transformation in water management policy of the Czech Republic. As a conclusion, water management policy served well for the aim of testing and validating the hypotheses of this research.

NOTES

1. Water resources refer to the fresh water available for use in a territory, and include surface waters (lakes, rivers, and streams) and groundwater. Freshwater availability in a country is primarily determined by climate conditions and inflows from other countries.
2. European Council Goteborg, Conclusions of the Presidency, June 15–16, 2001, Paragraph 19.
3. Decision No.1600/2002/EC of the European Parliament and of the Council of July 22, 2002, laying down the Sixth Community Environment Action Programme, L242/1.
4. The Council of the European Union Presidency, 10917/06, p. 13.
5. COM(2009) 147 final.
6. Ibid., p. 7.
7. Ibid., p. 12.
8. COM(2012) 673 final.

9. Ibid., p. 7–8.
10. Ibid., p. 8.
11. Ibid., final.
12. Ibid., 79 final.
13. Online source: http://ec.europa.eu/environment/newprg/index.htm (accessed on January 20, 2020).
14. Communication from the European Commission (2010) A Strategy for Smart, Sustainable and Inclusive Growth, COM(2010).
15. The Strategic Framework for Sustainable Development of the Czech Republic, p. 45.
16. Ibid., p. 46.
17. Ministry of Environment, State Environmental Policy of the Czech Republic (2012–2020), p. 6.
18. Ibid., p. 8.
19. Ibid., p. 8.
20. Interview with the former Minister of Environment of the Czech Republic, Bedrich Moldan,February 27, 2020, Prague.
21. Ibid.
22. Based on Government Resolution No.1452 of November 30, 2009.
23. Strategy on Adaptation to Climate Change in the Czech Republic, Government Resolution No.861 dated October 26, 2015.
24. Office of the Government of the Czech Republic, Department of Sustainable Development, *Strategic Framework Czech Republic 2030*, adopted by Government Resolution No.292 on April 19, 2017.
25. Interview with the former Minister of Environment of the Czech Republic.
26. The Plan of Main River Basins of the Czech Republic, Government Resolution No.562 (2007), p. 15–28.
27. European Commission SWD(2012) 379 final.
28. European Commission, SWD(2019) 35 final, p. 10.
29. Ibid., p. 14.
30. Ibid., p. 15.
31. Ibid., p. 21–22.
32. Act No. 156/1998 on Fertilisers, Supplementary Soil Substances, Supplementary Plant Preparations and Substrates and on Agrochemical Testing of Agricultural Land, June 12, 1998.
33. Government Regulation No. 103/2003 on the designation of vulnerable zones and on the use and storage of fertilizers and livestock manure, crop rotation, and implementation of erosion-control measures in these areas, March 3, 2003.
34. Amended by Government Regulation No.401/2015 of December 14, 2015, on indicators and values of permissible pollution of surface water and wastewater, requirements for permits for the discharge of wastewater into surface waters and sewers and on sensitive areas.
35. Directive 91/271/EEC rules that BOD_5 value can be 25 mg/l and COD_{Cr} value can be 125 mg/l in all categories of population.

36. The Czech Republic sets the ranges for BOD_5 in between 30 and 60 mg/l (500–2,000 persons), 25 and 50 mg/l (2,001–10,000 persons), 20 and 40 mg/l (10,001–100,000 persons), and 15 and 30 mg/l (over 100,000 persons); and the ranges for COD_{Cr} in between 125 and 180 mg/l (500–2,000 persons), 120 and 170 mg/l (2,001–10,000 persons), 90 and 130 mg/l (10,001–100,000 persons), and 75 and 125 mg/l (over 100,000 persons).

37. Government Regulation No. 401/2015 on indicators and values of permissible pollution of surface waters and wastewaters, requirements for permits for the discharge of wastewaters into surface waters and sewers and on sensitive areas, December 14, 2015.

38. Interview with the former Minister of Environment of the Czech Republic.

39. Government Decree No. 5/2011 on the delimitation of hydrogeological zones and groundwater bodies, the method of assessing groundwater status and the requirements of groundwater detection and evaluation programs, December 20, 2010.

40. European Commission, SWD(2019) 35 final, p. 113.

41. Act No. 258/2000 on the protection of public health, July 14, 2000.

42. Government Decree No. 252/2004 laying down hygiene requirements for drinking and hot water and the frequency and extent of drinking water inspections, April 22, 2004, as amended by Government Decree No.187/2005, May 4, 2005.

43. Directive 2000/60/EC establishing a framework for Community action in the field of water policy, Official Journal of the European Communities, L 327/1.

44. European Environment Agency (EEA), European Waters: Assessment of Status and Pressures 2018, Report No 7/2018, p. 25.

45. Ibid., p. 26.

46. European Commission Working Document, Second River Basin Management Plans—Czech Republic, SWD(2019) 35 final, 26.2.2019, p. 51

47. Ibid., p. 52.

48. Ibid., p. 59.

49. European Environment Agency (EEA), European Waters, p. 35.

50. European Commission Working Document, Second River Basin Management Plans, pp. 34–35.

51. Surface Water for Drinking Directive (75/440/EEC) sets standards for the BOD_5 value of surface water intended for the abstraction of drinking water. The level should not exceed 3 mg O2/l if the water is intended for abstraction with rapid filtration and disinfection only. A subsequent Drinking Water Directive (98/83/EC) does not set guideline or mandatory values of BOD_5 but defines indicator limits for ammonium: remedial actions for improving drinking water quality should be taken if an indicator value of 388 μg N/l is exceeded.

52. Drinking Water Directive (98/83/EC) sets maximum allowable concentration for nitrate as 50 mg/l. Surface Water for Drinking Directive (75/440/EEC) sets guideline concentration for nitrate as 25 mg/l.

54. Urban Waste Water Treatment Directive (91/71/EEC) aims to decrease organic pollution and involves the removal of nutrients.

55. European Environment Agency (EEA), Indicator Assessment, Published December 3, 2019. Online source: https://www.eea.europa.eu/data-and-maps/indic

ators/oxygen-consuming-substances-in-rivers/oxygen-consuming-substances-in-rivers-9 and https://www.eea.europa.eu/data-and-maps/indicators/nutrients-in-freshwater/nutrients-in-freshwater-assessment-published-9 (Accessed on January 20, 2020)

55. Ministry of Environment, Report on the Environment of the Czech Republic 2018, p. 70.

56. Directive 2008/105/EC of the European Parliament and of the Council of December 16, 2008, on environmental quality standards in the field of water policy, *Official Journal of the European Union*, L 348/84.

57. European Environment Agency (EEA), European Waters, p. 39.

58. Ibid., p. 42.

59. European Commission Working Document, Second River Basin Management Plans, p. 66.

60. European Environment Agency (EEA), European Waters: Assessment of Status and Pressures 2018, Report No 7/2018, p.44.

61. European Commission Working Document, Second River Basin Management Plans, p. 74.

62. European Environment Agency (EEA), European Waters, p. 57.

63. European Commission Working Document, Second River Basin Management Plans, p. 81.

64. Ministry of Environment, Report on the Environment of the Czech Republic 2018, p. 67.

65 European Environment Agency (EEA), European Waters, p. 53.

66. Ministry of Agriculture, Report on Water Management in the Czech Republic, 2011, 2012, 2015, 2016, 2017, 2018, as compared.

67. Ministry of Environment, Report on the Environment of the Czech Republic 2018, p. 81.

68. Ministry of Agriculture, Report on Water Management 2018, p. 61.

69. Eurostat, Water Statistics, Population connected to public water supply. Online code: env_wat_pop (Accessed on July 20, 2020).

70. Eurostat, Water Statistics, Total freshwater abstraction by public water supply. Online code: env_wat_abs (Accessed on July 20, 2020).

71. Ministry of Environment, Report on the Environment of the Czech Republic 2018, p. 68.

72. The two microbiological parameters Escherichia coli and Enterococci should be absent from drinking water to guarantee its quality and safety. Chemicals include trace elements, like arsenic, nickel, or lead, and other substances like cyanide or polycyclic aromatic hydrocarbon or nitrogen components—nitrate and nitrite. The indicator parameters have indirect relevance to the quality of water: they serve to indicate that something has changed in the source water, the treatment or the distribution of the water.

73. In the Czech Republic, data on drinking water quality is maintained by the Ministry of Health based on Decree No. 252/2004 which is harmonized with the European Council Directive 98/83/EC, on the quality of water intended for human consumption. This system includes all public water supplies.

74. Overview of the Drinking Water Quality in Czech Republic (2011–2013), p. 7.

75. 3.5 million analyses for microbiological parameters, 5.7 million analyses for chemical parameters, and 13.8 million analyses for indicator parameters were carried out.

76. European Commission (COM) 2016, 666 final, p. 7; and also European Topic Centre on Inland, Coastal and Marine Waters (2015) "Results of the reporting 2011-2013 under the Drinking Water Directive 98/83/EC," p. 8.

77. CENIA, online source: https://issar.cenia.cz/en/core-set-of-environmental-indicators/water-management-and-water-quality/waste-water-treatment/ (Accessed on February 10, 2020).

78. Czech Statistical Office (CZSO). Online source: https://www.czso.cz/documents/10180/91605333/2800211909.pdf/7ff82ef5-fcae-4c28-8a92-dfaf48eee56a?version=1.0 (Accessed on June 22, 2020).

79. CENIA, online source: https://issar.cenia.cz/en/core-set-of-environmental-indicators/water-management-and-water-quality/waste-water-treatment/ (Accessed on February 10, 2020).

80. Eurostat, share of the population connected to at least secondary urban wastewater treatment. Online code: env_ww_con (Accessed on July 27, 2020).

81. Ministry of Agriculture and Ministry of Environment, Report on Water Management in CR (2018), p. 33.

82. Ministry of Agriculture and Ministry of Environment, p. 33.

83. Houses of Parliament, Parliamentary Office of Science and Technology, Diffuse Pollution of Water by Agriculture, Postnote No.478, October 2014.

84. Ministry of Agriculture, the Ecophyto 2018 Plan, September 10, 2008.

85. Margaux Lapierre, Alexandre Sauquet, Subervie Julie (2019) Providing technical assistance to peer networks to reduce pesticide use in Europe: Evidence from the French Ecophyto plan, https://hal.archives-ouvertes.fr/hal-02190979v2

86. The Danish Eco-Innovation Programme, online source: https://eng.ecoinnovation.dk/media/mst/8051431/CASE_oerestad_artikel.pdf (Accessed on July 27, 2020).

87. Online source: http://www.hammarbysjostad.se/en/water-and-sewage/ (Accessed on July 27, 2020.

88. Ministry of Agriculture, Report on Water Management in 2018, p. 19.

89. European Commission, Environmental Implementation Review 2019, SWD (2019) 119 final, p. 19.

90. Ibid., p. 19.

91. Eurostat, Share of the population connected to at least secondary urban wastewater treatment, Online code: env_ww_con (Accessed on July 27, 2020).

92. The policy-maker preferred not to disclose his name and institutional affiliation.

93. European Commission, COM(2018) 337 final.

Chapter 4

Waste Management

INTRODUCTION: IMPORTANCE OF THE WASTE MANAGEMENT POLICY

Waste is an important environmental issue because nearly every human activity is associated with the production of waste. Since waste generation cannot be avoided, states and companies need to device ways of using or disposing waste in a manner that does not represent a threat to human health and to the environment. Good waste management by individuals, households, businesses, and local authorities helps to achieve a more resource-efficient society. Circular economy entails gradually decoupling economic activity from the consumption of finite resources. Thus, it provides the industry with valuable secondary raw materials, cutting down on the amount of material needed from the natural environment. It also presents an economic opportunity because recycling of waste creates an entire sector by changing the way we design products that can be made to be made again.

A country's effort for achieving environmental sustainability is reflected in that country's effectiveness in reusing and recycling its resources. With this aim, a proper waste policy should favor the practical application of a five-step waste hierarchy. The waste hierarchy ranks waste management options according to what is best for the environment. It gives top priority to preventing waste in the first place. When waste is created, then the proper sequence of waste management should be reuse, recycling, and recovery as much as feasible, and disposal should be the last resort.

In this framework, avoiding and reducing the generation of waste is the highest priority, and it is compatible with the concept of sustainable development and respecting its principles. It encourages the community, industry, and government to reduce the amount of virgin materials extracted and used.

The goal of this policy is to maximize efficiency and to avoid unnecessary consumption. In connection with this strategy, individuals should prefer avoiding disposable goods or single-use materials; selecting items with the least packaging or the ones that require the fewest resources to produce; buying products that are recycled, recyclable, repairable, refillable, reusable, or biodegradable. States, on the other hand, should promote the technological changes that reduce waste generation.

The second priority, that is, the resource recovery, focuses on maximizing the options for reuse, recycling, reprocessing, and energy recovery. This strategy is applied when avoiding and reducing waste is not possible. Reuse without further processing and recycling keep materials in the productive economy and benefit the environment by decreasing the need for new materials.

Finally, disposal is the last resort in the waste hierarchy. In such cases when the waste cannot be safely recycled, then disposal becomes the only viable option. It is the worst option for the environment because it signifies the loss of resources. The goal should be a progressive reduction of landfilling in order to prevent detrimental impacts to human health and environment.

The revolutionary changes became possible in European states by means of new policy norms and strategies toward circular economy. They created a new mentality that resources can be used in sustainable ways. Waste is now perceived as a resource to be exploited rather than as an environmental problem.

EUROPEAN AND DOMESTIC POLICY NORMS IN WASTE MANAGEMENT

EU requirements have been an important driver of improvement in waste management in the EU member states. The EU Strategy Papers and Programmes established community-wide objectives in waste management area, and they have been the initiator and driver of changes and developments in the waste management policies of the member states. In the first part of this analysis, we will analyze the European Union's and the Czech Republic's *executive portfolio in waste management* with an aim to find out the level of resonance of general policy norms, objectives, and strategies.

In June 2001, the European Council adopted the first *EU Sustainable Development Strategy* in Goteborg which stressed the significance of sustainable development as a fundamental objective, and of decoupling the economic growth from resource use. The presidency conclusions declared that "the relationship between economic growth, consumption of natural resources and the generation of waste must change. Strong economic performance must go hand in hand with sustainable use of natural resources and levels of waste."[1]

In 2006, same principles were adopted in the context of the *Renewed EU Sustainable Development Strategy*. This Strategy Paper aimed to present a single, coherent strategy on how the EU will more effectively meet its commitment to the principles of sustainable development. The Strategy Paper identified an operational objective in the area of conservation and management of natural resources as "avoiding the generation of waste and enhancing efficient use of natural resources by applying the concept of life-cycle thinking and promoting re-use and recycling."[2]

This theme was further developed in the Sixth Environment Action Programme,[3] which presented a program of Community action on environment with key objectives covering a period of ten years. The priorities of the Sixth EAP were climate-change, nature, and biodiversity, health and quality of life, natural resources, and waste. It developed a vision integrating resource, product, and waste policies by calling for the development of seven thematic strategies including a strategy on the sustainable use of natural resources and a strategy on the recycling of waste. Regarding waste management, the program aimed to achieve "better resource efficiency as well as better resource and waste management to bring about more sustainable production and consumption patterns, thereby decoupling the use of resources and the generation of waste from the rate of economic growth to ensure that the consumption of renewable and non-renewable resources does not exceed the carrying capacity of the environment" (Article 2.2). The document strongly emphasized the need for a strategic approach to meet these environmental objectives (Article 3), promoted development of new Community legislation and amendment of existing legislation (Article 3.1), and more effective implementation and enforcement of Community legislation in member states (Article 3.2).

As it is impossible to develop a waste policy without assessing the wider impacts, the scope of recycling strategy was later expanded to cover both prevention and recycling of waste. Towards a Thematic Strategy on the Prevention and Recycling of Waste,[4] adopted by the European Commission in 2003, enlarged the scope of the strategy to cover all scopes of the waste management. This strategy was based on two premises: waste policy should focus on the environmental impact of using resources, and waste policy should take a life-cycle approach.

Thematic Strategy on the Prevention and Recycling of Waste,[5] adopted in 2005, set the long-term goal for the EU to become a recycling society that seeks to avoid waste and uses waste as a resource. The Strategy Paper outlined seven key areas through which the EU can move toward improved waste management. To achieve these objectives, policy strategies were delineated as "more and better recycling, less waste going to landfills, and more compost and energy recovery from waste."[6] These strategies were expected to lead to significant environmental, economic, and social benefits.

The review of the Thematic Strategy was carried in 2010. Thematic Strategy on the Prevention and Recycling of Waste[7] assessed the progress made with implementing the actions proposed by the Thematic Strategy 2005: improving implementation and enforcement of the existing EU waste legislation at the national level; simplification and modernization of the EU waste legislation; introduction of the life-cycle thinking in waste policy and waste prevention; and improving the knowledge base, development of recycling standards, and further elaboration of the EU's recycling policy.

The reviewed strategy document assessed the progress made in the achievement of the long-term objectives of the Thematic Strategy 2005, and concluded that the Thematic Strategy 2005 had played an important role in guiding policy developments in the EU: "Significant progress has been achieved in the improvement and simplification of legislation, establishment and diffusion of key concepts such as waste hierarchy and life-cycle thinking, setting focus on waste prevention, coordination of efforts to improve knowledge, and setting new European collection and recycling targets."[8] As a result, it stated, "recycling rates have improved, the amount of waste going to landfill has decreased, and environmental impacts have decreased"[9] notwithstanding that there were large implementation and enforcement differences between member states.

The Seventh Environment Action Programme,[10] which entered into force in January 2014, is the guiding European environment policy strategy for the period up to December 31, 2020. It set out the long-term objectives for the sustainable use of resources by stating: "Our prosperity and healthy environment stem from an innovative, circular economy where nothing is wasted, natural resources are managed sustainably; biodiversity is protected, valued and restored in ways that enhance our society's resilience."[11] Within the program, there was a special focus on turning waste into a resource with more prevention, reuse, and recycling, as well as on phasing out wasteful and damaging practices like landfilling.

The program has the following priority objectives: "to turn the Union into a resource-efficient, green and competitive low-carbon economy; to safeguard the Union's citizens from environment-related pressures and risks to health and well-being; to maximize the benefits of Union environment legislation by improving implementation" (Article 2.1). To attain these objectives, the program is based on the precautionary principle, the principles of preventive action and of rectification of pollution at source and the polluter-pays principle (Article 2.2). All measures, actions, and targets set out in the program are subject to a comprehensive impact assessment (Article 2.4).

After becoming a member of the European Union, the Czech Republic adopted these policy norms and strategies. The Strategic Framework for Sustainable Development of the Czech Republic (2010) presented its waste management policy objective as promoting sustainable materials management. To achieve this goal, it defined the policy strategies as:

> providing support for increasing the proportion of environmentally friendly low-waste technologies and BAT; supporting research, development, and innovation in environmentally sound and knowledge technologies with high added value and lower material consumption requirements; applying a system for minimizing, sorting and subsequently using waste for material recovery; raising consumer awareness on sustainable consumption and production; promoting education and awareness on sustainable consumption and production.[12]

State Environmental Policy of the Czech Republic for the period 2012–2020 sets the framework for an effective protection of environment in the Czech Republic. For sustainable use of resources, its policy objectives are prevention and minimization of waste, waste reuse, and waste recovery. The policy paper underlines that landfilling is a major issue in the Czech Republic and that landfilling is "the most important and urgent problem to be tackled in the short and medium-term period."[13] Regarding the prevention and minimization of waste, the State Environmental Policy reaffirms its compliance with the waste management hierarchy as defined in the EU legislation (first, prevention, then reuse, recycling, recovery, and finally, disposal), and defines its policy objectives in waste management as reduction of the share of landfilling on the total waste disposal processes, increase energy recovery of municipal and similar waste, and prevention of waste production.

Strategic Framework Czech Republic 2030[14] replaced the Strategic Framework of Sustainable Development 2010. The new document formulated strategic and specific objectives that the Czech Republic should fulfill by the year 2030. The strategies in waste management were grouped under two main objectives: waste prevention, and waste separation and recycling. The policy goal regarding waste prevention was to achieve better use of domestic resources by replacing natural materials with recycled waste and secondary raw materials. To enable this policy strategy, the state supports the waste management hierarchy, where waste prevention is preferred to recycling, recycling to energy recovery, and energy recovery of waste to waste disposal, referred to as circular economy.[15] Regarding waste separation and recycling, the policy strategy focused on the prohibition of the landfilling of usable and recyclable waste. Mixed communal waste, only after the process of waste separation, shall be used following the applicable legislation. Separation of biowaste should also become a standard practice for Czech cities and municipalities over the next fifteen years, and it should be used to produce biogas.[16]

Conclusive Remarks

In the previous section, I put forward that the EU-accession process produces two varieties of policy transfer: hard transfer of programs and their implementation, and soft transfer of ideas, concepts, and attitudes. In this section, I presented that after its accession to the EU, the Czech Republic transferred the ideas, concepts, and strategies presented in the waste management policy of the EU. As a result, national policy norms and strategies were formulated in accordance with the EU strategic documents. A comparative analysis of policy norms and objectives set by the EU and by the Czech Republic in the last two decades demonstrates and validates the theoretical hypothesis of the soft transfer of ideas, concepts, and attitudes from the supranational level to the national level. Next section analyzes the hard transfer of the policy legislation.

EUROPEAN AND DOMESTIC POLICY LEGISLATION IN WASTE MANAGEMENT

In this section, we will analyze the legal framework in waste management at the European level, and then, analyze how this supranational legislative development was transposed in the Czech Republic.

The EU member states began taking national measures to control and manage waste in the 1970s. The disparity between the provisions on waste disposal among the member states urged the need for a common waste management policy at the EU level. A common EU waste policy framework was established on July 15, 1975, with an aim to approximate national laws in this field. The *Waste Framework Directive (75/442/EEC)* intended to reconcile considerations that disparities between the provisions on waste disposal already applicable or in preparation in the various member states may create unequal conditions of competition, thus, negatively affect the functioning of the common market. The essential objective of the Directive was the protection of human health and environment against harmful effects caused by the collection, transport, treatment, and storage of waste. To achieve these policy objectives, it obliged the member states to take appropriate steps to encourage the prevention, recycling and processing of waste, the extraction of raw materials and energy from waste, and any other process for the reuse of waste (Article 3). It also obliged the member states to take the necessary measures to ensure that waste is disposed of without endangering human health and without harming the environment (Article 4). It required that the competent national authorities shall prepare one or several plans relating to the type and quantity of waste to be disposed of, general technical requirements for the disposal of waste, suitable disposal sites, and any special arrangements for particular wastes (Article 6). The Directive also introduced the polluter pays principle (Article 11).

Almost thirty years later, in its conclusions of July 1, 2004, the Council called on the Commission to bring forward a proposal for the revision of certain aspects of the WFD to differentiate clearly between recovery and disposal, and to clarify the distinction between waste and non-waste. The amendments to the WFD were provided by the *Directive 2008/98/EC*,[17] which stated its primary objectives as minimizing the negative effects of the generation and management of waste on human health and the environment, reducing the use of resources, and favoring the practical application of the waste hierarchy (Preamble Article 6). The amended Directive arose from certain needs to modify the concepts. First of all, definitions of waste recovery and waste disposal were needed to be clarified based on genuine difference in their environmental impact through the substitution of natural resources in the economy and recognizing the potential benefits to the environment and human health of using waste as a resource (Preamble Article 19). Thus, the amended Directive set the basic definitions related to waste management, such as definitions of waste, recycling, and recovery in Article 3. There was also a need to differentiate between the various aspects of the waste definition, such as byproducts that are not waste and waste that ceases to be waste (Preamble Article 22). So, the Directive explained when waste ceases to be waste and becomes a secondary raw material (end-of-waste status) in Article 6. The Directive provided measures on waste prevention such as polluter pays principle (Article 14) and the extended producer responsibility (Article 15). By means of these measures, the Directive aimed to move the EU closer to a recycling society, seeking to avoid waste generation and using waste as a resource (Preamble 28). It also incorporated provisions on hazardous waste (Articles 17–20), waste oils (Article 21), and biowaste (Article 22).

The Directive included two new recycling and recovery targets to be achieved by 2020:

> by 2020, the preparing for re-use and the recycling of waste materials such as at least paper, metal, plastic and glass from households and possibly from other origins shall be increased to a minimum of overall 50% by weight; and preparing for re-use, recycling and other material recovery, including backfilling operations using waste to substitute other materials shall be increased to a minimum of 70% by weight. (Article 11)

Finally, the Directive obliged member states to prepare waste management plans and waste prevention programs to be integrated either into the waste management plans or into other environmental policy programs. National waste management plans shall set out an analysis of the current waste management situation in the geographical entity concerned, as well as the measures to be taken to improve environmentally sound preparing for reuse, recycling, recovery and disposal of waste and an evaluation of how the plan will support the implementation (Article 28). National waste prevention programs shall set out the waste prevention objectives, and describe the existing prevention measures and evaluate the usefulness of the measures with an aim to break the link between economic growth and the environmental impacts associated with the generation of waste (Article 29). Both the WMPs and the WPPs should be submitted to the Commission (Article 33) and be evaluated at least every sixth year (Article 30).

Directive 2018/851/EU[18] is the latest amendment to the WFD. The revised WFD aims to improve the quantity and the quality of the resources fit for reuse and recycling by fostering separate collection of waste. The revised WFD introduced new obligations for the member states regarding the separate collection of waste:[19] Waste shall be subject to separate collection and shall not be mixed with other waste or other materials with different properties (Article 10.2); derogations for the separate collection obligations can apply (Article 10.3); member states are obliged to collect separately at least paper, metal, plastic, and glass, and member states shall also set up separate collection for textiles (Article 11.1); member states shall set up separate collection for hazardous waste fractions produced by households (Article 20); member states shall ensure that biowaste is either separated and recycled at source, or is collected separately (Article 22.1). Separate collection is obligatory as of 2015 for paper, metal, plastic, and glass, by December 31, 2023, for biowaste, and by January 1, 2025, for textile and hazardous household waste. By December 31, 2021, member states shall submit a report to the Commission on the implementation of the requirements on separate collection.

The revised WFD also introduced ambitious targets. While the average recycling rate for municipal waste in EU average in 2017 was 46 percent, the revised WFD increased the targets for reuse and recycling of waste to 55 percent, 60 percent, and 65 percent by 2025, 2030, and 2035, respectively (Article 11.2). Thus, substantial efforts will be needed across Europe to transpose this Directive and achieve the targets successfully. Member states are required to bring into force the laws, regulations, and administrative provisions necessary to comply with this Directive by July 5, 2020.

Council Regulation EEC/259/93 on the supervision and control of shipments of waste within, into, and out of the European Community implemented the Basel Convention on the Control of Transboundary Movements of Hazardous Wastes and Their Disposal into the EC. The Regulation laid down the procedures in the shipment of waste for disposal (Articles 3–5) and for recovery (Articles 6–11) from one member state to another member state. It prohibited all exports of waste for disposal, except those to EFTA countries (Article 14.1); and prohibited all exports of waste for recovery, except those countries which are parties to the Basel Convention and/or with which the Community or its member states have concluded bilateral, multilateral, or regional agreements (Article 16.1). It prohibited all imports of waste for disposal into the Community except those from EFTA countries and from countries which are parties to the Basel Convention and/or with which the Community or its member states have concluded bilateral, multilateral, or regional agreements (Articles 19–20), and prohibited all imports of waste for recovery into the Community except those countries which are parties to the Basel Convention and/or with which the Community or its member states have concluded bilateral, multilateral, or regional agreements (Articles 21–22).

The EU introduced the first measures on the management of packaging waste with the Directive 85/339/EEC, which set rules on the production, marketing, use, recycling, and refilling of containers of liquids for human consumption and on the disposal of the used containers. In the following years, some member states started introducing their own measures to address the environmental aspects of packaging and packaging waste. As a consequence, diverging national legislation appeared, and the situation called for harmonization at the European level. To harmonize national measures concerning the management of packaging and packaging waste and to prevent or reduce their impact on the environment, the EU adopted the Packaging Directive[20] (Directive 1994/62/EC). The policy objective of the Directive was to harmonize national measures concerning the management of packaging and packaging waste and to provide a high level of environmental protection (Article 1). To this end, it laid down the measures aimed at preventing the production of packaging waste (Article 4), reusing (Article 5), and recycling and recovering packaging waste (Articles 5–6). It encouraged the reuse systems of packaging in an environmentally sound manner (Article 5).

In order to comply with the policy objectives of this Directive, member states were obliged to adopt national programs or similar actions (Article 4). They were also obliged to take the necessary measures to attain the following targets: (a) achieve between 50 percent and 65 percent by weight of the packaging waste to be recovered no later than five years; and (b) achieve between 25 percent and 45 percent by weight of the packaging material contained in packaging waste to be recycled (Article 6). Member states were also obliged to take the necessary measures to ensure the return and collection of packaging waste from the consumers, final users, or from waste stream in order to channel it to the most appropriate waste management alternatives (Article 7). Finally, member states were obliged to include in their waste management plans a specific chapter on the management of packaging and packaging waste (Article 14).

In 2004, this Directive was amended to provide criteria clarifying the definition of the term "packaging," and increase the targets for recovery and recycling of packaging waste. In 2005, the Directive was revised again to grant new member states transitional periods for attaining the recovery and recycling targets. The latest revision was made on April 29, 2015, with the adoption of Directive 2015/720/EU on the consumption of lightweight plastic carrier bags.

Directive 1999/31/EC on the Landfill of Waste[21] aimed to introduce operational and technical requirements on the waste and landfills, provide for measures, procedures, and guidance to prevent or reduce negative effects of the landfilling of waste on the environment and on the human health as far as possible (Article 1). The Directive obliged member states to set up a national strategy for the implementation of the reduction of biodegradable municipal waste (BDMW) going to landfills not later than two years after the entry into force of this Directive (Article 5). This strategy should ensure that (a) not later than five years after the entry into force of this Directive, BDMW going to landfills must be reduced to 75 percent of the total amount of BDMW produced in 1995; (b) not later than eight years after the entry into force of this Directive, BDMW going to landfills must be reduced to 50 percent of the total amount of BDMW produced in 1995; (c) not later than fifteen years after the entry into force of this Directive, BDMW going to landfills must be reduced to 35 percent of the total amount of BDMW produced in 1995.

Member states were also obliged to grant permit for landfills based on an impact assessment of the proposed operation plan of the disposal site which complies with all the relevant requirements of this Directive (Articles 7–8). For the existing landfills, member states were obliged to take measures such as the preparation and submission of conditioning plan by the operator of the landfill for the approval of the competent authorities (Article 14). Regarding the implementation of this Directive, European Commission obliged the member states to report at intervals of three years (Article 15).

This Directive was amended by the Directive 2018/850/EU.[22] The policy objective of the new Directive was to ensure a progressive reduction of landfilling of biodegradable waste by strengthening landfill restrictions through stringent operational and technical requirements on the waste and landfills, and by providing measures, procedures, and guidance to prevent or reduce the negative effects of landfilling of waste on environment and on the human health (Article 1). This amendment was necessary because biodegradable municipal waste still accounts for a large proportion of municipal waste, and in this way, poses significant negative environmental effects in terms of greenhouse gas emissions and pollution of surface water, groundwater, soil, and air (Preamble Article 7).

This Directive added new provisions regarding the target of landfilling: First of all, member states shall endeavor to ensure that as of 2030, all waste suitable for recycling or other recovery, in particular in municipal waste, shall not be accepted in a landfill (Article 4.c). Moreover, they shall take the necessary measures to ensure that by 2035 the amount of municipal waste landfill is reduced to 10 percent or less of the total amount of municipal waste generated (Article 4.5). Attainment of this target might be postponed by a member state up to five years based on the condition that member state landfilled more than 60 percent of its municipal waste generated in 2013 (Article 4.6).

Directive 2000/76/EC on the incineration of waste[23] was adopted with an aim to prevent or to limit the negative effects on the environment of the pollution by emissions into air, soil, surface water, and groundwater, and the resulting risks to human health from the incineration of waste (Article 1). The Directive required any incineration plant to operate only with a permit granted by the competent authority (Article 4.1). The operator of the incineration plant should take all necessary precautions concerning the delivery and reception of waste in order to prevent or limit the negative effects on environment and human health (Article 5.1). The Directive required incineration plants to be designed, equipped, built, and operated in such a way that the emission limit values set out in the Annex of this Directive are not exceeded in the exhaust gas (Article 7.1). Regarding the residues resulting from the operation of the plant, the Directive required these residues to be minimized and recycled, where appropriate, in accordance with the Community legislation (Article 9).

In the Czech Republic, the main act of transposing the requirements of the WFD into the domestic law is the Waste Act (Act No. 185/2001). This act incorporated the relevant EU legislation on waste prevention and management regarding environmental protection, human health protection, and sustainable development with an aim to reduce the adverse impacts of the use of natural resources and to improve the efficiency of such use (Article 1). It provided the definition of waste concepts (Articles 3–4), rights and obligations of persons in waste management (Article 16), and responsibilities of the public authorities in waste management (Article 17).

Regarding one of the most important objectives of waste management, that is, separate collection of waste, the Act stated that the originator and authorized person shall be obliged to classify the waste according to the waste catalog, which shall be issued by the Ministry of the Environment by implementing legal regulation (Article 5.1). The Directive defined the conditions when waste is classified as hazardous waste (Article 6) and the evaluations of hazardous waste properties (Articles 7–8).

The Waste Act of the Czech Republic incorporated another important EU norm on waste management, that is, waste hierarchy, stating that in the context of waste management, the following waste hierarchy must be respected: (a) waste prevention, (b) preparation for reuse; (c) waste recycling, (d) other uses of waste such as energy recovery, (e) waste disposal (Article 9).

Another important EU policy norm, that is, waste prevention, was transposed by stipulating that everyone has, in the course of his activity or within the scope of his competence, an obligation to prevent waste generation, to limit its quantity and hazardous properties; wastes whose formation cannot be prevented must be recovered or disposed of in a manner that does not endanger human health and the environment and which is in compliance with this Act and with special legal regulations (Article 10).

The Act provided provisions for landfilling of waste. It stipulated the obligations of the operator of the waste landfill (Article 21.1), the location and technical implementation of the waste landfill (Article 21.2), and list of wastes prohibited from being landfilled (Article 21.5). The Act prohibited storing mixed municipal waste and recyclable and recoverable wastes on landfills from 2024 onward (Article 21.7). Regarding the incineration of waste, the Act stipulated that waste may be incinerated only if the conditions laid down in the legislation on air protection and on energy management are met (Article 22).

Regarding the waste management and waste prevention programs, the Act transposed the EU requirement by providing that the waste management plan shall be prepared for the purpose of creating conditions for the prevention and management of waste (Article 41.2). According to this provision, waste prevention program shall include the waste prevention objectives, waste prevention measures, and a set of indicators to evaluate the achievement of waste prevention objectives and measures, which will enable monitoring and assessment of the progress made through them (Article 41.a.2). The binding part of the Waste Management Plan of the Czech Republic shall contain the objectives and measures for waste prevention, and set out the objectives, principles, and measures for their achievement including (a) municipal waste management, in particular mixed municipal waste and biodegradable waste, (b) management of construction waste, (c) packaging waste management, (d) disposal of end-of-life products, (e) management of other wastes, in particular hazardous wastes, (f) preparation for reuse, recycling, recovery and disposal of waste, minimizing adverse environmental impacts, (g) reducing the amount of waste landfilled, in particular biodegradable waste, and (h) reducing the proportion of biodegradable component in mixed municipal waste (Article 42.5).

The preparation of the Waste Management Plan is assigned to the MoE (Article 42). Regional administrations are also assigned to prepare waste management plan of the region (Article 43). Municipalities are also assigned to prepare municipal waste management plans for the waste they produce and the waste they manage on the condition that if they produce more than 10 tons of hazardous waste or more than 1,000 tons of other waste annually (Article 44.1).

In the Czech Republic, the export, import, and transit of wastes is regulated by the Act No. 125/1997 (Act on Waste). This act transposed both the EU and international legislation[24] into domestic law. It introduced a control system consisting of color lists where the consent of the MoE is required for import, export, and transit of wastes for recovery specified in Red and Amber Lists of wastes and wastes not specified in any of these lists. Wastes destined for recovery that are included in the Green List are allowed to move across the frontiers between the OECD countries as freely tradable commodities.

Act No. 185/2001 (Waste Act) also provided the transposition of the EU Directive on shipment of waste. Part 9 of the Waste Act regulated the cross-border waste transport,[25] and prohibited the transboundary shipment of wastes to the Czech Republic for the purpose of their disposal, with the exemption of waste arisen in neighboring states as a consequence of natural disasters or states of emergency (Article 54.2). The competent authority for the transboundary shipment of waste to the Czech Republic was set as the MoE. On transboundary shipment of waste, MoE can decide on granting consent or objection in accordance with the regulation of the EU on shipment of waste (Article 56).

Act No. 477/2001 (Packaging Act) aimed to protect the environment by preventing the generation of wastes from packages by reducing the weight, volume, and hazards of the packages and chemical substances contained in such packaging in compliance with European Community law (Article 1.1). This Act laid down the rights and obligations of self-employed legal and natural persons (Article 5) and the competence of administrative authorities (Articles 31–36) in the handling of packaging waste, and the associated protective measures (Article 43). The Act incorporated one of the EU's waste policy norms, that is, the prevention of waste, into national legislation by stating that the person who places packaging on the market shall ensure that the weight and volume of the packaging are kept to a minimum in order to reduce packaging waste (Article 3.1). The Act also stipulated that the plastic carrier bags can only be provided to the consumer for the reimbursement of costs corresponding to the cost of its acquisition (Article 3.3). The Act incorporated another EU waste policy norm, that is, the waste hierarchy, in the national legislation by stipulating that the packaging after its intended use shall be reusable or it shall be converted into a product or raw material (recycling), or subject to direct combustion for the release of energy by combustion, either alone or together with other waste (energy recovery), or biodegradable constituents of this waste shall be subject to anaerobic treatment under controlled conditions (organic recycling) (Article 4).

Regarding waste separation, Decree No. 381/2001[26] established a waste catalog and stipulated that the waste generators and authorized persons shall classify wastes as six-digit catalog numbers as listed in the waste catalog, in which the first two digits denote the waste group, the second two digits denote the subgroup, and the third two digits denote the waste type (Article 2.1). The Decree obliged the producers of municipal waste to sort this waste according to the Waste Catalog in accordance with the system set by the municipality (Article 2.3). Furthermore, Act No. 229/2014[27] made separate collection of recyclable municipal waste (paper, plastics, glass, metals, biowaste) mandatory in all municipalities, and obliged municipalities to provide locations for storing biodegradable municipal waste (Article 3). Decree No. 321/2014 on the extent and method of ensuring separate collection of municipal waste components further specified the methods of ensuring separate collection of municipal waste components, that is, BDMW (Article 2), paper, plastic, glass, and metals (Article 3), and hazardous municipal waste (Article 4). Act No. 223/2015 provided an amendment to the Waste Act. It defined separate collection as collection where the waste stream is separated according to the type, category, and nature of the waste in order to facilitate specific treatment (Article 8). It also required a procedure for the hazardous components of waste to be assessed by one or more authorized persons on the basis of a request from the producer or an authorized person, and to submit an application through an integrated system in compliance with environmental reporting obligations (Article 16).

The Czech Republic adopted the waste management plan (WMP)[28] on July 1, 2003, to establish a sustainable waste management policy. The plan was drawn up for a period of ten years, that is, from 2003 to 2012. The structure and content of the plan followed Articles 41 and 42 of the Act on Waste and Article 28 of the WFD of the EU. Policy objectives of the WMP were closely linked to the prevention and minimization of waste. In order to attain these policy objectives, the binding part of the WMP stipulated the creation of normative, economic, and administrative instruments. The plan had the main goal of transposing the EU Directives concerned with waste management into the legal regulations of the Czech Republic. With this goal, it stipulated preparation of regional waste management plans; creating fees to be paid by the waste generator (in accordance with the polluter pays principle); creating a financial reserve for recovery and decontamination of landfills; creating deposits for returnable packages; creating fines on misdemeanors; supporting desirable activities leading to the prevention of waste generation; and preferring products from recycled materials and environmentally friendly products in public procurement by public administrative bodies at all levels. It also established a waste management council as an inter-sectoral advisory body of the MoE for the coordination of waste management planning at the national level.[29]

Article 29 of the Directive 2008/98/EC required member states to create national programs on waste prevention by December 12, 2013. Waste Act of the Czech Republic (Act No. 185/2001 Coll.) also stipulated this obligation in its Article 42. As a consequence, waste prevention programme (WPP)[30] was prepared by the MoE with participation of the relevant stakeholders at the level of ministries, professional associations, and representatives of business, research institutions, and NGOs. This fulfills the requirement set by Article 31 of the Directive 2008/98/EC, which requires member states to ensure that relevant stakeholders and authorities and the general public have the opportunity to participate in the elaboration of the waste management plans and waste prevention programs. On December 12, 2013, the draft program was published on the Ministry's website. The program was approved by the government on October 27, 2014 by Government Resolution No. 869/2014. The WPP focused on the reduction of the amount of waste produced and on the reduction of its hazardous properties that have an adverse impact on environment and public health.

The WPP analyzed and monitored the following waste streams: municipal waste, biodegradable waste, food waste, end-of-life products (packaging, electrical, battery, accumulators, end-of-life vehicles), construction waste, textile waste/textile for reuse. The policy objectives of the WPP were "to support introduction of low-waste and waste-free innovative technologies that will conserve input of raw materials; boost the use of secondary materials; re-use end-of-life materials and products; pay maximum attention to food waste and create conditions for gradual reduction of the volume at all levels of the food cycle; promote returnable reusable packaging; implement environmental education and awareness."[31]

The Government of the Czech Republic also adopted WMP for 2015–2024 with Regulation No. 352/2014 on December 22, 2014, in order to fulfill the mandatory EU target set out in the Article 28 of Directive 2008/98/EC and in the Articles 41 and 42 of the Waste Act. During preparation, the Plan was repeatedly consulted and reviewed by the public, and it was discussed with key representatives from the waste management sector. The plan came into force on January 1, 2015, for the period of ten years, that is, the period 2015–2024. WMP stated its policy objectives as preventing waste; applying the waste management hierarchy; quality recycling and maximum recovery of suitable waste (material, energy, biological); optimizing the management of biodegradable municipal waste; mandatory introduction of separate collection for waste; energy recovery from municipal waste; substantial reduction of landfilling.[32]

In *municipal waste*, WMP CR defined its policy objectives as: (a) by 2015, introduce separate collection at least for waste consisting of paper, plastics, glass, and metals, and (b) by 2020, increase to at least 50 percent by weight preparing for reuse and recycling at least for waste consisting of paper, plastic, metal, and glass (Article 3.3.1.1).[33]

In *biodegradable waste*, WMP CR defined its specific policy objective as: by 2020, reduce the maximum quantity of biodegradable municipal waste deposited at landfills to maximum 35 percent by weight of the total quantity of biodegradable municipal waste produced in 1995 (Article 3.3.1.3).

In *separate collection of waste,* WMP CR put forward its policy strategies as "ensuring separate collection of paper, metal, plastic, glass, bulky waste, hazardous components of municipal waste; separate collection of recoverable components of municipal waste through sufficient and accessible collection points in communities; separate collection of bio-waste; separate collection of hazardous components of municipal waste; free acceptance of end-of-life products from citizens in take-back locations."[34]

In *packaging waste*, WMP CR defined its policy objectives by 2020 as (Article 3.3.1.6.1):

a. "increase the overall packaging recycling to the level of 70%;
b. increase the overall recovery of packaging waste to the level of 80%;
c. increase the recycling of plastic packaging to the level of 50%;
d. achieve 55% overall recovery of consumer sale packaging;
e. achieve 50% recycling of consumer sales packaging."[35]

WMP CR is the defining document for the development of waste management plans of the regions as well. Individual regions in the Czech Republic are obliged to prepare their own strategies keeping the binding part of WMP as the mandatory basis. This strategy aims to ensure consistency between national WMP and each individual regional WMPs. It is also possible for the municipalities to deviate from the waste hierarchy but only in justified cases in compliance with applicable legislation and if this does not endanger or harm the environment or human health.

Conclusive Remarks

This legislative analysis demonstrated that the Czech Republic has transferred the EU legislation on waste management into its domestic structure. Theoretical implication of this empirical analysis is that it validated the research hypothesis on policy transfer mechanisms. I put forward earlier that

Table 4.1 Comparison of the Policy Legislation in the EU and in the Czech Republic (legislative adaptation)

	European Union's Legislative Portfolio	*Czech Republic's Legislative Portfolio*
General Framework	Directive 75/442/EEC Directive 2008/98/EC Directive 2018/851/EU	Act No. 185/2001
Separate Collection of Waste	Directive 2008/98/EC Directive 2018/851/EU	Decree No. 381/2001 Act No. 229/2014 Decree No. 321/2014 Act No. 223/2015 Waste Management Plan of the Czech Republic (2015-2024)
Recycling of Waste	Directive 2008/98/EC	Waste Management Plan of the Czech Republic (2015-2024)
Landfilling of Waste	Directive 1999/31/EC Directive 2018/850/EU	Waste Management Plan of the Czech Republic (2015-2024)
Shipment of Waste	Regulation EEC/259/93	Act No. 125/1997 Act No. 185/2001
Packaging Waste	Directive 85/339/EEC Directive 94/62/EC Directive 2015/720/EU	Act No. 477/2001 Waste Management Plan of the Czech Republic (2015-2024)

Source: Author's own compilation

the soft transfer emerges when member states transfer the ideas, concepts, and strategies presented in the EU's policy documents, as opposed to the hard transfer that emerges when the EU transfers its rules and procedures to member states. While the first mechanism is related to the concepts of social learning or ideational learning, the second one is conceptualized as a mandatory requirement. The first part of this empirical research focused on the soft transfer through a content analysis of the executive portfolio of the EU and Czech Republic in waste management policy, and concluded that the normative adaptation is evident. The second part of this empirical research focused on the hard transfer through a content analysis of the EU legislation and domestic legislation in waste management in a comparative perspective, and reached to the conclusion that legislative adaptation is also evident (see table 4.1). These findings signify a domestic structural change in waste management policy.

IMPLEMENTATION OF THE POLICY NORMS AND LEGISLATION IN WASTE MANAGEMENT

This part aims to analyze the implementation of the new policy norms and policy legislation in the Czech Republic. As discussed before, what this book offers as a contribution to the existing literature on Europeanization is to empirically test and validate the completeness of policy transfer through a statistical data analysis based on a set of indicators. Positive changes in these indicators will reveal successful operationalization of the normative and legislative adaptation, and therefore, fulfillment of the policy objectives and strategies in waste management.

The set of indicators under consideration in this part is categorized into four areas:

1. Overall waste generation
   - Production of waste per capita
   - Production of waste by sectors
2. Municipal waste generation
3. Waste separation
4. Waste treatment

Regrettably, other indicators had to be discarded from the date set for the sake of clarity and efficiency of this research. When using the statistical data, a mix of ex-ante (factual) and ex-post (using only available data) approach is adopted. The positive changes in indicators reflect the level of achievement of the operationalization of the policy norms and strategies in waste management.

The statistical data used here comes from the national and European statistics. The Czech Republic's two information systems for waste generation and treatment are the Waste Management Information System, managed by the Czech Environmental Information Agency (CENIA) on behalf of the Ministry of Environment; and the waste statistics managed by the Czech Statistical Office (CSO). However, their data differ in terms of definitions, reporting boundaries and surveying methods. So, they are complemented by other data provided by Eurostat, European Environment Agency, and OECD. This list of statistical information is used here with an aim to analyze and compare the detectable and empirically provable domestic changes in this policy area. The main principle here is that the data chosen for comparison is obtained from periodic surveys or other sources regularly providing reliable data in the time series.

Overall Waste Generation

I start this section with the empirical analysis of the reduction of overall waste generation in the Czech Republic and in EU average in comparison. This analysis provides a comparative data set that signifies the progress associated with this waste management policy objective both in the Czech Republic and in the EU.

When we analyze the waste generated per person in the Czech Republic in the time period 2004–2016, we see that annual generation of waste per capita decreased from 1,736 kg in 2004—at the time of EU accession—to 1,214 kg in 2016, corresponding to a 43 percent decrease. There are various trends in other member states: In the time period 2004–2016, we see an increasing trend in the generation of waste per capita in Poland (27%), Ireland (25%), Germany (22%), and Italy (20%). On the contrary, we see a decreasing trend in the generation of waste per capita in Romania (74%), Finland (73%), Hungary (67%), Portugal (59%), and Austria (51%) in the same time period. Therefore, there is not a unique trend for this indicator in the EU-zone.[36]

The amount of waste generated by different sectors shows striking differences among the EU member states. For instance, while in Czech Republic, construction and demolition waste makes up 40 percent of the total waste generation, higher figures are found in Germany (55%), Denmark (58%), France (69%), the Netherlands (69%), Austria (73%), and Luxembourg (75 percent). Mining and quarrying make up the biggest share of waste generation in Bulgaria (81%), Romania (86%), and Greece (78%) whereas in Belgium, Denmark, Latvia, the Netherlands, Slovenia, and Austria it is around 0.1 percent. Manufacturing has a higher percentage in Lithuania (41%), Estonia (36%), and Ireland (34%).[37,38]

As a comparison, when we analyze the amount of waste generated by different sectors in the Czech Republic in the time period 2006–2018, we find that (a) construction and manufacturing sectors are the main waste generators; (b) increase in the waste generated by the construction sector can be attributed to the development of the Czech economy and the increased trend in urbanization; and (c) there has been a significant decrease in the waste generated by the mining sector and the electricity, gas, steam, and air condition sectors.[39]

Municipal Waste Generation

In the period from 2004 to 2018, the municipal waste generation per capita in the Czech Republic showed a steady increase. Statistics show that in 2004—at the time of EU accession—GDP per capita in the Czech Republic was 12.840 Euro and generation of municipal waste per capita was 279 kg. These figures have steadily increased from 2004 to 2018. In 2018, GDP per capita was 17.980 Euro and generation of municipal waste per capita was 351 kg. Thus, from 2004 to 2018, GDP recorded a 40 percent increase and municipal waste generation recorded a 25 percent increase. Based on these figures, the increase in generation of municipal waste can be attributed to the increased household consumption as a result of economic development.[40]

Statistical data on the municipal waste generation per capita in the Czech Republic and in other EU member states shows different tendencies in 2005 and in 2018 (see figure 4.1). When we compare the figures by only using the exact data available and neglecting the estimated ones, we reach to these conclusions: A decrease in the generation of municipal waste per capita is observed in Sweden, Poland, Slovakia, Romania, Finland, Latvia, Estonia, and Slovenia. On the other hand, an increase in the generation of waste per capita is observed in Denmark, Austria, Estonia, Czechia, Lithuania, and Bulgaria. These variations reflect differences in consumption patterns and economic wealth, but also depend on how municipal waste is collected and managed[41] (see figure 4.2).

Separate Collection

Article 10.2 of the WFD (Directive 2018/851) stipulated that waste shall be subject to separate collection and shall not be mixed with other waste or other materials with different properties. Regarding the implementation of this legislation, the Czech Republic took the necessary measures: in most municipalities, separate containers are provided for the collection of paper, plastics, and glass, and a small progress has been recorded. While the separate collection of paper, glass, plastics, and metals accounted for 14 percent

	Mining and quarrying	Manufacturing	Energy	Construction and demolition	Other economic activities	Households
EU-27	27.6	11.1	3.4	34.8	14.7	8.3
Belgium	0.1	23.3	1.2	31.0	36.4	8.0
Bulgaria	81.9	2.9	7.9	1.7	3.2	2.4
Czechia	0.6	18.4	3.5	40.0	23.5	14.1
Denmark	0.1	4.8	3.9	58.3	16.3	16.6
Germany	1.8	14.0	2.5	55.1	17.2	9.4
Estonia	26.0	36.6	25.1	4.8	5.8	1.8
Ireland	15.7	34.7	2.1	10.0	27.6	9.9
Greece	78.4	6.4	3.5	0.8	4.2	6.6
Spain	15.7	11.1	3.0	27.8	25.6	16.8
France	0.7	6.8	0.4	69.4	13.6	9.0
Croatia	12.0	8.4	2.3	24.5	31.0	21.7
Italy	0.5	17.0	1.6	33.3	29.3	18.4
Cyprus	5.3	33.0	0.1	35.6	10.0	16.0
Latvia	0.0	19.4	11.4	4.4	30.4	34.4
Lithuania	0.6	41.2	2.0	7.6	31.7	16.8
Luxembourg	0.4	7.5	0.1	75.2	10.6	6.3
Hungary	0.9	16.9	16.2	22.5	25.3	18.2
Malta	7.7	1.4	0.1	68.9	13.4	8.4
Netherlands	0.0	9.8	1.5	69.9	12.8	6.1
Austria	0.1	9.2	0.8	73.4	9.5	7.0
Poland	38.8	16.5	11.3	10.4	17.8	5.2
Portugal	2.8	17.3	0.6	11.6	34.5	33.2
Romania	86.7	4.4	3.9	0.2	2.6	2.3
Slovenia	0.2	26.0	14.3	9.9	38.0	11.5
Slovakia	3.0	32.5	9.0	9.1	28.6	17.8
Finland	76.2	7.6	0.9	11.3	2.6	1.5
Sweden	77.5	4.1	1.3	6.9	7.1	3.1

Figure 4.1 Waste Generation by Economic Activities and Households, 2016. *Source*: Eurostat, Statistics Explained, Waste Statistics, p.2, Online code: env_wasgen (accessed on July 27, 2020).

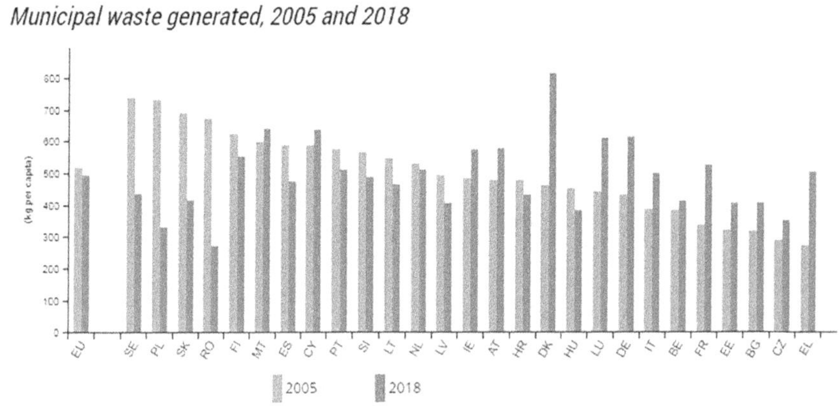

Figure 4.2 Municipal Waste Generated in EU Member States in 2005 and 2018. *Source*: Eurostat.

in 2012, it increased to 16 percent in 2018. The standard waste collection accounted for 68 percent in 2012 and this amount decreased to 56 percent in 2018.[42]

Article 42 of the Preamble of the WFD (Directive 2018/851) highlights that separate collection can be achieved via a range of collection arrangements such as through door-to-door collection, bring and reception systems, or other collection arrangements. When we analyze the collection systems used in the member states, we find that most EU member states use door-to-door collection, co-mingled door-to-door collection, bring points, and civic amenity sites. A detailed analysis of different approaches to collection arrangements gives these conclusions: (a) paper is mainly collected within door-to-door systems in fourteen member states; (b) glass is mainly collected within bring systems in eighteen member states; (c) plastic is collected within door-to-door systems in eighteen member states, of which only four collect plastic as a separate fraction and others apply co-mingling with other fractions; (d) metal is collected within door-to-door systems by seventeen member states while the Czech Republic, Slovakia, and Latvia collect metal only at civic amenities; and (e) biowaste is separately collected door to door in fourteen member states, while other member states do not collect biowaste separately in their primary collection system[43,44] (see table 4.2).

The separation of waste is part of daily habits of citizens which changes very slowly. The municipalities should determine their collection system with the future in mind and keep it stable while providing citizens clear instructions and awareness campaigns.

Table 4.2 Overview of Municipal Waste Collection Systems in the EU Member States

Collection type	*Paper*	*Glass*	*Plastic*	*Metal*	*Bio-waste*
Door-to-door (single fraction)	AT, BE, BG, CY, DE, DK, FI, HU, IT, LU, LV, NL, SI, UK	BG, FI, LU, LV, NL, SI, MT	AT, LV, NL, DK	FI, NL, DK	AT, BE, CZ, DE, FI, EE, IT, HU, LU, NL, SI, SE, IE, UK
Co-mingled (plastic+metal)			BE, BG, CY, DE, FR, IT, HU, LU, SI		
3 fractions	RO, MT: paper, plastic, metal UK: plastic, metal, glass				
All in one bin	EL, IE				
Bring points	CZ, EE, ES, FR, HR, LT, PT, PL, SE, SK	AT, BE, DK, CY, CZ, DE, EE, ES, FR, HR, IT, HU, LT, PT, PL, RO, SE, SK	SE ES, HR, LT, PT, PL	AT, EE, SE	ES
Civic amenity sites	CZ (metal), SK (metal and bio-waste), LV (metal) Addition collection of all waste streams: all countries				

Source: European Commission (2015) Assessment of separate collection schemes in the 28 capitals of the EU, Reference: 070201/ENV/2014/691401/SFRA/A2

Treatment of Waste

There are significant differences regarding the waste treatment methods used in the EU member states. In the EU-27 average in 2016, slightly more than a half (52.6%) of the waste was treated in recovery operations: recycling (36.7% of the total treated waste), backfilling (10.1%), or energy recovery (5.8%). The remaining 47.4 percent was either landfilled (40.3%), incinerated without energy recovery (0.8%), or disposed of otherwise (6.3 percent). There are significant differences observed among the EU member states concerning the use of various waste treatment methods. For instance, some member states had very high recycling rates (Italy, Belgium, and Latvia), while others favor landfill (Greece, Bulgaria, Romania, Finland, and Sweden).[45] Article 11.2 of the revised WFD (Directive 2018/851/EU) increased the targets for reuse and recycling of waste to 55 percent, 60 percent, and 65 percent by 2025, 2030, and 2035, respectively. Statistics show that in 2016, Italy, Belgium, and Latvia were already in compliance with this article (see figure 4.3).[46]

Regarding waste treatment, major EU policy norm is the reduction of landfilling. In the time period 1995–2018, total municipal waste landfilled in the EU fell by 69 million tons, or 57 percent—from 121 million tons (286 kg per capita) in 1995 to 52 million tons (117 kg per capita) in 2018. This

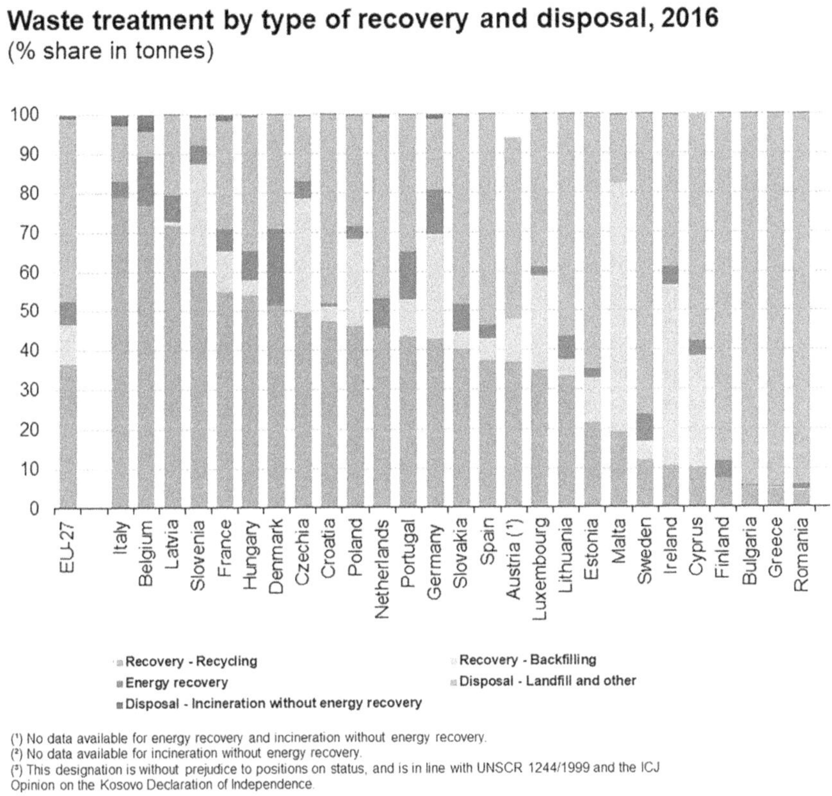

Figure 4.3 Waste Treatment by Type of Recovery and Disposal in EU Member States, 2016. *Source*: Eurostat, Statistics Explained, Waste Statistics, p.8. Online code: env_wastrt (accessed July 27, 2020).

corresponds to an average annual decline of 3.5 percent. For the shorter period 2005–2018 landfilling has fallen by as much as 3.7 percent per year on average. As a result, the rate of landfilled waste as share of generated waste in the EU dropped from 61 percent in 1995 to 24 percent in 2018. On the other hand, the amount of waste recycled (material recycling and composting) rose from 37 million tons (87 kg per capita) in 1995 to 104 million tons (233 kg per capita) in 2018 at an average annual rate of 4.2 percent. The share of municipal waste recycled overall rose from 19 percent in 1995 to 47 percent in 2018.[47]

The reduction in landfilling and the increase in recycling in the EU-zone can partly be attributed to the implementation of European legislation that adopted ambitious targets for the recycling of municipal and packaging waste and lower limits for landfill of municipal waste[48] (see figure 4.4.).

In comparison with the EU average, statistical data for the Czech Republic shows that the percentage of landfilling of the municipal waste has greatly

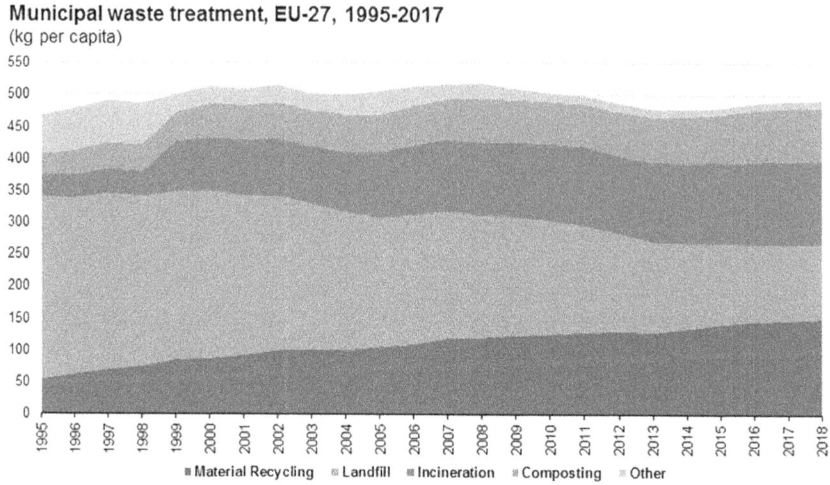

Figure 4.4 Municipal Waste Treatment in EU-27 (1995–2017). *Source*: Eurostat, Statistics Explained.

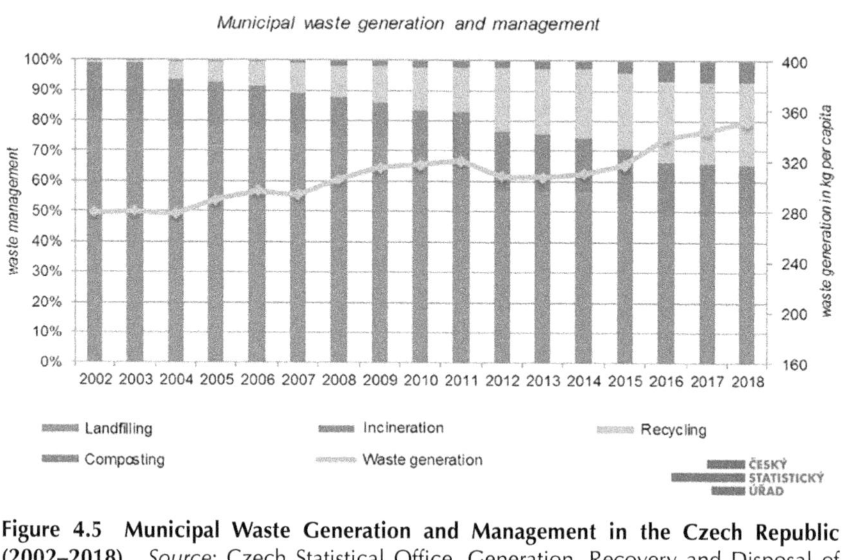

Figure 4.5 Municipal Waste Generation and Management in the Czech Republic (2002–2018). *Source*: Czech Statistical Office, Generation, Recovery and Disposal of Waste 2018, Code: 280020-19 (accessed on July 27, 2020).

diminished from more than 80 percent in 2002 to 49 percent in 2018. Another successful policy implementation comes with the recycling rate, which rose from almost undetectable amount in 2002 to 27 percent in 2018[49] (see figure 4.5).

Article 4.5 of the Directive 2018/850/EU stipulated that by 2035 the amount of municipal waste landfilled is to be reduced to 10 percent or less of the total amount of municipal waste generated. However, in the Czech Republic, landfilling of the municipal waste was still the main treatment method in 2018 with a 49 percent share, which was followed by recycling method with a 27 percent share.[50]

This analysis demonstrated that the problem of landfilling persisted in the Czech Republic because it has been the cheapest way to dispose of the waste. It is preferred due to its low cost; however, it does not allow adequate development of the circular economy, competing technologies, and the creation of jobs. Ministry of Environment attempted to overcome this problem through the Waste Act revision. A draft proposal was prepared, and after the interministerial and stakeholder discussion, it passed through the Government Legislative Council in summer 2016. However, landfillers intensively lobbied against the new law. On the media, it was debated if the higher landfilling fee means a higher financial burden on the citizens.[51] The debate was primarily on the implementation of the new Directive (2018/850/EU) which introduced new tools to meet ongoing requirements for municipal waste separation: If municipalities are able to gradually separate from 35 percent in 2019 to 75 percent in 2027 of their municipal waste, they will pay less for the rest of the waste they could possibly landfill. The debate on media and in society together with the lack of time before the parliamentary election in 2017 resulted in the freezing of the adoption of the new waste legislation. An environmental NGO in the Czech Republic, Hnuti Duha, argued that the increase in the landfilling fee that is supplemented by a recycling discount for municipalities is a positive development.[52] Bedrich Moldan, former Minister of Environment, stated that there are three key issues jeopardizing the implementation of the waste management policy strategies: "First of all, people need to accept the new provisions and pay more for their waste. Secondly, economy of the waste management is difficult to manage without state intervention and the system can't work properly without state subsidies. Third issue is about the technology; the Czech Republic is not efficient in recycling technology yet."[53] The Confederation of Industry of the Czech Republic also argued that "there is practically no recycling industry in the Czech Republic in the field of municipal waste; and by applying the circular economy approach, a significant business opportunity will arise for domestic companies."[54]

Best Practices for Waste Management

An assessment of the European Commission of the separate collection schemes in the capital cities of the EU-28 member states revealed that Ljubljana, Tallinn, Dublin, Helsinki, Vienna, and Luxembourg are the best examples in the EU-zone with the highest separate collection rates for five fractions combined (paper, metal, glass, plastic, and biowaste) measured as a percentage of the estimated generation of these fractions. An important note here is that this statistical data is only provided for the capital cities, and it may not reflect the overall performance of the individual member states[55] (see figure 4.6).

The best performing capital cities provide some of the best available techniques for a better municipal waste management. Although these cities apply different collection schemes and have different rates of development, it appears that they have some common traits:

> First of all, they have Pay As You Throw system which charges households according to the amount of waste they generate. This system is one of the main success factor for successful separate collection of waste fractions. Secondly, the municipality and the producer responsibility schemes or free market mechanisms for recyclables are combined smartly and in a harmonized way. Thirdly, primary success factor for the collection of bio-waste is the establishment of two measures, i.e. the fee system and the municipal regulations setting the minimum standard for separate collection. Finally, communication to households is very clear about what can and what cannot be placed in each bin. The successful examples achieved to cultivate an interest in the general population about how waste should be properly managed.[56]

We can analyze the best available techniques offered by some of the most successful examples. The best performer in this area is the city of Ljubljana. A key ingredient for Ljubljana's successful result was the introduction of door-to-door collection, which was the largest contributor to the sharp increase in recycling rates.[57] Residents of multi-apartment buildings and houses use common bins for residual waste, biological waste, packaging, and paper, while those living in family houses usually have their own bins for separate waste collection.[58] As separate collection increased, the amount of residual waste constantly declined. The public waste management company, Snaga, also uses social media to improve the user friendliness of the collection services. Use of social media and SMS communication on collection dates are important factors in achieving the high capture rate. Moreover, the underground collection units in the city center facilitate collection without visual nuisance.[59]

Another best performer, Luxembourg, applied a legal enforcement method to achieve the separate collection policy objective. The Luxembourg waste management law of March 21, 2012, obliged apartment buildings to have separate waste collection facilities. In addition to the classic valuable substances like paper, glass, and packaging, biowaste should also be collected separately in apartment buildings in addition to various hazardous substances as medication, lighting appliances, electric appliances, batteries, packaging with paint residues or other hazardous substances, and spray cans.

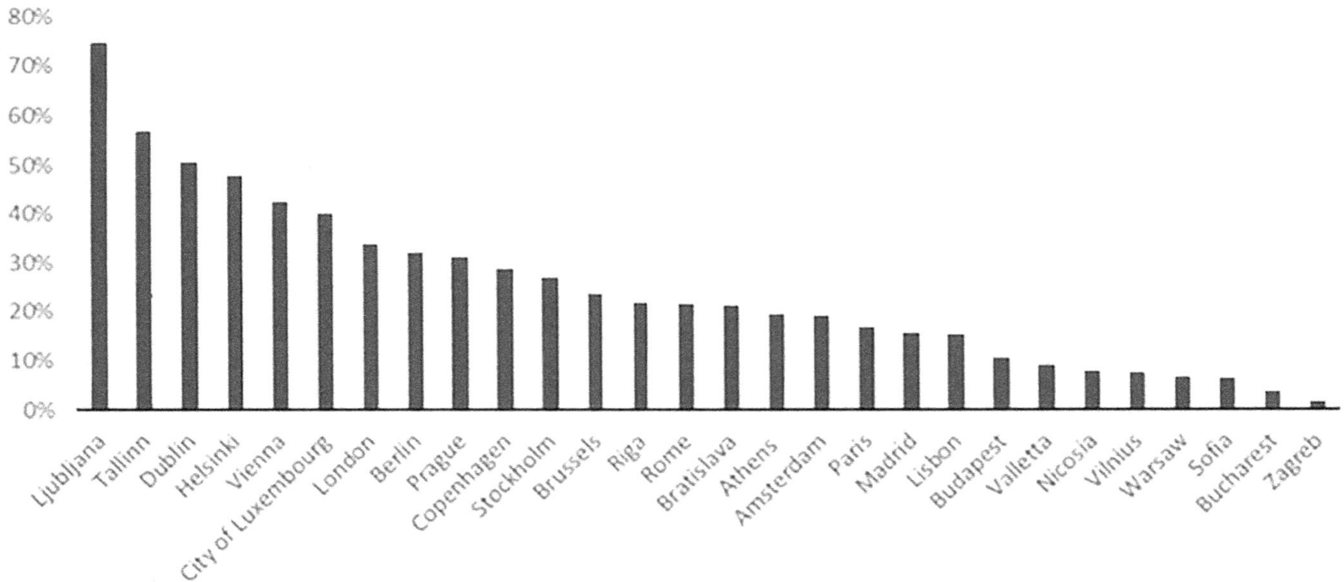

Figure 4.6 Combined Rate for Separate Collection of Paper, Metal, Glass, Plastic and Biowaste for EU-28 Capitals. *Source*: European Commission (2015) Assessment of separate collection schemes in the twenty-eight capitals of the EU, Reference: 070201/ENV/2014/691401/SFRA/A2.

The building management organizes the transport of these products. The integrated waste collection system called SuperDrecksKëscht provides free advisory services for the building managers to support local implementation: onsite visit for analysis of the existing situation, instruction/training of the property management to use the collection station, briefing of inhabitants, training of building management staff, and label certification. A quality label, SuperDrecksKëscht®, is also introduced for environmentally friendly waste management. It promotes businesses and institutions from the private and public sector that make active contribution to environmental protection by implementing a modern waste management system.[60] This approach has contributed to high separate collection rates in Luxembourg.

Another best practice for the reduction of landfilling is from the UK where a landfill tax was introduced in 1996 as an environmental tax to help reduce the amount of waste going to landfill. The landfill tax has steadily risen as an incentive to reduce landfill and encourage recycling. The tax is split into two bands. A lower band is applied to inert waste that contains no biodegradable material, while all other general waste has the standard band applied. The tax is charged by weight and currently the standard rate is 94.15 £/ton and the lower rate is 3 £/ton, effective as of April 1, 2020.[61] This economic incentive has transformed waste management in the UK toward landfill diversion and recycling.[62]

For the management of municipal waste, Eco Island Krk project provides another best practice. The project was introduced in June 2005 on the island of Krk in Croatia. The project is an ecologically based system for the management of municipal waste, which is the first of its kind in Croatia that treats all types of waste. In total, 6,500 containers are installed in 1,400 places on the island in order to ensure the successful operation of the scheme. Domestic waste is collected in five separate containers (organic waste, paper, PET waste, glass, and other). The island of Krk also has seven POSAMs, which are special collection areas (each local autonomy unit has one) where citizens may drop off larger household waste, old furniture, white goods, old cars, car tyres, and dangerous waste free of charge. In 2015, the municipality separated 50 percent of its waste and prepared it for reuse and recycling, thereby already meeting the 2020 target set in the Waste Framework Directive. Great importance is placed on the promotion of the system and education of users.[63]

CONCLUSIVE REMARKS

Empirical Conclusions

In this chapter, we have seen significant adaptation of the Czech Republic's normative and legal structure to that of the EU in the waste management area.

Regarding *normative adaptation*, the comparative analysis demonstrated that the EU policy norms and policy strategies have been significantly transposed into the Czech Republic's national strategies and action programs. To be precise, resource efficiency and the principles of a sustainable material economy and of a reduced environmental impact of material flows are enshrined in the State Environmental Policy and in the National Strategy for Sustainable Development. The main objectives were set as preventing waste and moving toward a circular economy, minimizing the negative health and environmental impact of waste management, maximizing waste reuse and recovery, and ensuring raw material supply security, all of which are in accordance with the EU policy norms. The policy objectives of the current WMP (2015–2024) include increasing reuse and recovery and reducing the landfill of waste. This overview signifies normative adaptation to the EU policy norms in waste management.

Regarding *legislative adaptation*, the Czech Republic undertook to incorporate the requirements stipulated in the EU regulations into the national legislation. In waste management legislation, it has come a long way since its waste act in 2001. Since then, transposition of the EU Directives has been a continuous process. According to the OECD evaluations, the country has successfully aligned its policy framework with the EU legislation.[64] The current legislation in waste management fully respects the requirements of the relevant EU Directives, and the legal framework is fairly complete with quantitative targets. Therefore, one can conclude that waste management policy in the Czech Republic has undergone considerable changes in the post EU-accession phase, and put a system in place that generally corresponds to the current requirements of European legislation.

This study underlines the significance of the implementation of policy objectives and policy legislation that is acquired as a result of policy transfer. We argue that the extent of the operationalization of policy objectives, strategies, and legislation constitutes the final indicator of the domestic change. Based on this approach, an empirical analysis of the implementation phase is carried out based on national and international statistics. This analysis provides the completeness of policy transfer.

The implementation of these policy objectives and policy legislation in the Czech Republic showed positive achievements, however, a number of implementation problems are also observed. First of all, rate of recycling is low (27%) in relation to the EU target: by 2020, the preparing for reuse and the recycling shall be increased to a minimum of overall 50 percent by weight (Article 11 of Directive 2008/98/EC). The corresponding rate for the EU-27 average is 47 percent in 2018. Secondly, the Czech Republic has made progress in waste recovery but landfilling remains the main treatment method. Landfilling, though reduced, remains the most common waste treatment method with a rate of 49 percent in 2018, which is very high compared to the EU-27 average of 24 percent in 2018. Finally, there is inadequate waste prevention. These problems prevent the full implementation of the EU waste management policy and the achievement of relevant policy objectives.

A major obstacle for policy analysts is to monitor progress in waste management in the Czech Republic due to the existence of two information systems that produce data that differ greatly. The Czech Statistical Office and the Ministry of Environment produce different data and therefore give different messages. The availability of reliable statistical data is critical for improving assessment of Czech waste policies.

As a future trajectory, current plans and programs on waste management and prevention should help the waste management policy of the Czech Republic in moving forward. The potential for progress is good.

Theoretical Implications

These empirical findings points out significant structural change in waste management policy of the Czech Republic. This section briefly summarizes the theoretical implications of these empirical findings for each of the research hypotheses.

In the bureaucratic organization, we witnessed establishment of new institutions dealing with environment and waste management as a result of post-Communist transition, but not as a result of EU accession. After the EU accession, existing institutional structure continued in the waste management because it was perceived as adequate to handle the adoption of the EU's acquis. Thus, we don't observe bureaucratic reorganization in the waste management area as a result of EU accession. However, there has been an obvious change in the administrative procedures in the environmental policy in general, and in the waste management policy in particular. This change has taken place through the adoption and implementation of EU's methods for assessment, information, planning, auditing, and sanctioning as discussed in the previous section. This section demonstrated first a discernable normative adaptation of the Czech Republic's waste policy norms, policy objectives and strategies to those of the EU, and secondly, a complete transposition of the contents of the EU Directives in waste management. Regarding the operationalization of such systemic changes, this chapter also showed that these newly adopted policy norms, objectives, strategies, and legislation have been implemented extensively in the Czech Republic, although with some shortcomings and gaps. All these empirical findings validate a structural transformation in waste management policy of the Czech Republic. As a conclusion, waste management policy served well for the aim of testing and validating the hypotheses of this research.

NOTES

1. European Council Goteborg, Conclusions of the Presidency, June 15–16, 2001, Paragraph 31.
2. The Council of the European Presidency Union, 10917/06, p. 13.
3. Decision No 1600/2002/EC of July 22, 2002.
4. COM(2003) 301 final.
5. COM(2005) 666 final.
6. Ibid.
7. COM(2011) 13 final.
8. Ibid., p. 8.
9. Ibid.
10. Decision No. 1386/2013/EU of November 20, 2013, on a General Union Environment Action Programme to 2020 "Living well, within the limits of our planet."
11. Ibid., Article 1 of Annex.
12. The Strategic Framework for Sustainable Development of the Czech Republic, p. 32.
13. Ministry of Environment, State Environmental Policy of the Czech Republic (2012–2020), p. 5.
14. Office of the Government of the Czech Republic, Department of Sustainable Development, Strategic Framework Czech Republic 2030, adopted by Government Resolution No.292 on April 19, 2017.
15. Ibid., p. 58.
16. Ibid., p. 89.
17. Directive 2008/98/EC of November 19, 2008, Official Journal L312/4.
18. Directive 2018/851/EU of May 30, 2018, Official Journal L 150/109.
19. European Commission, Guidance for separate collection of municipal waste, April 2020.
20. Directive 94/62/EC of December 20, 1994, Official Journal L365/10.
21. Directive 1999/31/EC of April 26, 1999, Official Journal L182/1.
22. Directive 2018/850/EU of May 30, 2018, Official Journal L150/100.
23. Directive 2000/76/EC of December 4, 2000, Official Journal L332/91.
24. Directive EEC/259/93, through which EC implemented the Basel Convention on the Control of Transboundary Movements of Hazardous Wastes and Their Disposal, and also Decision of the OECD Council 92/39/Final, on the control of transfrontier movements of wastes destined for recovery operations.
25. This part was also complimented by the Decree No. 381/2001 Coll. establishing a Waste Catalog, List of Hazardous Wastes, and List of Waste and States for the Purposes of Export, Import and Transit of Waste.
26. Decree No.381/2001 of the Ministry of the Environment establishing the Waste Catalog, October 17, 2001.
27. Act No.229/2014 amending Act No.185/2001 on waste and amending certain other acts, September 23, 2014.

28. The binding part of the Plan was announced by Government Order No.197/2003 and amended by the Government Order No.473/2009 Coll. The Government Order No.181/2013 Coll. extended the Plan until December 31, 2014.

29. Waste Management Plan of the Czech Republic, Regulation No. 197/2003 Coll., p. 40–42.

30. According to the Waste Framework Directive, Waste Prevention Programs can either be included in waste management plans or may be developed separately. Act No. 154/2010 Coll. incorporated the waste prevention program into the Waste Management Plan of the Czech Republic.

31. Waste Management Plan of the Czech Republic, Ministry of Environment, Prague, November 2014, p. 139–140.

32. Ibid., p. 90–91.

33. Ibid., p. 104.

34. Ibid., p. 135.

35. Ibid., p. 114.

36. Eurostat, Online data code: T2020_RT100, Last update: 31.01.2020 (accessed on July 27, 2020).

37. Eurostat, Statistics Explained, Waste Statistics, p. 2, Online code: env_wasgen.

38. Ibid.

39. Czech Statistical Office, Generation, Recovery and Disposal of Waste 2018, Online Code: 280020-19, Waste generated by enterprises: by selected economic activity (2006–2018) (accessed on July 27, 2020).

40. Eurostat, Municipal waste by waste management operations, Online code: env_wasmun, Accessed on 27.7.2020; Eurostat, Real GDP per capita. Online code: online data code: SDG_08_10 (accessed on July 27, 2020).

41. Eurostat, Waste Statistics, Online code: env_wasmun (accessed on July 27, 2020).

42. Czech Statistical Office, Generation, Recovery and Disposal of Waste 2012, Code: e-2001-13, Municipal waste by type of collection (accessed on July 27, 2020); and Czech Statistical Office, Generation, Recovery and Disposal of Waste 2018, Code: 280020-19, Municipal waste by type of collection (accessed on July 27, 2020).

43. European Commission (2015) Assessment of separate collection schemes in the twenty-eight capitals of the EU, Reference: 070201/ENV/2014/691401/SFRA/A2.

44. European Commission (2015) Assessment of separate collection schemes in the twenty-eight capitals of the EU, Reference: 070201/ENV/2014/691401/SFRA/A2.

45. Eurostat, Statistics Explained, Waste Statistics, p. 7, Online code: env_wastrt

46. Ibid.

47. Eurostat, Statistics Explained, Online source: https://ec.europa.eu/eurostat/statistics-explained/index.php/Municipal_waste_statistics (accessed on July 27, 2020).

48. Eurostat, Energy, Transport and Environment Indicators 2018, p. 156.

49. Czech Statistical Office, Generation, Recovery and Disposal of Waste 2018, Code: 280020-19, Municipal waste generation and management (2002–2018) (accessed on July 27, 2020).

50. Czech Statistical Office, Generation, Recovery and Disposal of Waste 2018, Online Code: 280020-19, Municipal Waste Management (accessed on July 27, 2020).
51. Online source: http://4liberty.eu/the-czech-republic-needs-new-waste-legislation (accessed on January 1, 2020).
52. Online source: https://ekolist.cz/cz/zpravodajstvi/zpravy/na-uterni-vladu-miri-nova-odpadova-legislativa-jsou-k-ni-vyhrady (accessed on January 1, 2020).
53. Interview with the former Ministry of Environment of the Czech Republic, Bedrich Moldan, February 27, 2020, Prague.
54. Online source: https://www.spcr.cz/aktivity/z-hospodarske-politiky/12237-pohled-prumyslu-na-odklon-od-skladkovani (accessed on January 1, 2020).
55. European Commission (2015) Assessment of separate collection schemes in the twenty-eight capitals of the EU, Reference: 070201/ENV/2014/691401/SFRA/A2.
56. European Commission (2015) Assessment of separate collection schemes in the twenty-eight capitals of the EU, Reference: 070201/ENV/2014/691401/SFRA/A2, p. 18.
57. Snaga, Waste Management Services, Online source: https://www.vokasnaga.si/en/separating%20waste (accessed on February 27, 2020).
58. Snaga, Collection Sites, Online source: https://www.vokasnaga.si/en/separating-and-collecting-waste/collection-sites-and-snagas-bins (accessed on July 27, 2020).
59. Snaga, Underground Collection Units, Online source: https://www.vokasnaga.si/en/separating-and-collecting-waste/underground-collection-units, (accessed on February 27, 2020).
60. SuperDrecksKëscht, Online source: https://www.sdk.lu/index.php/en/reverse-consumption/ecological-waste-management-in-residences/the-concept (accessed on July 27, 2020).
61. HM Revenue and Customs, Online source: https://www.gov.uk/government/publications/rates-and-allowances-landfill-tax/landfill-tax-rates-from-1-april-2013 (accessed on July 27, 2020).
62. Online source: https://www.360environmental.co.uk/legislation/waste_legislation/landfill_tax/ (accessed on July 27, 2020).
63. Online source: http://www.krk.hr/en/the_island_of_Krk/Eco_island_Krk (accessed on July 27, 2020).
64. OECD Environmental Performance Reviews: Czech Republic 2018, p. 148.

Chapter 5

Conclusion

How to preserve the quality and quantity of world's water sources and how to separate, treat, and recycle our municipal waste so that they won't pose negative impacts on the environment and on human and animal health are perhaps the most pressing issues of our decade.

In this book, I focused on the water and waste management policy areas and analyzed the impact of EU policy norms, objectives, strategies, and legislation on the Czech Republic's policy-making and policy implementation in the post-EU-accession phase. I started this analysis by defining Europeanization as the emergence and development of distinct rules and procedures and structures of governance at the European level, and tried to demonstrate how the ongoing process of European-level policy development has impacted on the domestic institutions and domestic policy-making in a member state. The research was confined only to the public policy, not political structures, or structures of representation. The study of public policies in this book contained the study of content of policies, legal and administrative structures that have been established to oversee them, and the dominant style in which policy is made and implemented. So, the approaches originating from a political perspective were neglected here.

This book covered the environmental policy because this policy area has recorded a significant development since it first emerged in the 1970s. In the 1980s, the first wave of laws had been passed in the more environmentally advanced countries. Moreover, various international organizations had launched environmental programs for the first time. As a result, while some countries were progressively developing their own environmental policies, others lagged behind. Given the transboundary character of most environmental problems, the EU sought to harmonize environmental standards among its members. In the 1990s, comprehensive environmental policies had been put in place in the EU level. This development also influenced the Central and Eastern European states that encountered serious environmental problems during the Communist rule. Association Agreements required some movement toward EU environmental standards, and eventual EU membership in 2004 required full application of Community environmental legislation in the CEEs. As a result, policy transfer is acquired in environmental policy and domestic structures adopted to the EU policy rules, regulations, methods, and strategies.

While analyzing domestic policy change as an impact of the development of supranational policy norms, objectives, and legislation, there were two analytical problems: The first problem was related to the measurement of changes in domestic policy, while the second one was related to detecting the EU's influence as compared to other possible causes of domestic policy change. Any study which attempts to examine the interlinked phenomena of Europeanization and domestic structural change must confront these issues. I solved the first problem by including the policy implementation as a measure of policy change, and the second problem by making a content analysis—it demonstrated how much of the EU regulations and directives was transferred into the Czech legislation. The comprehensive legislative development rendered the environmental policy a particularly appropriate area for the investigation of this impact.

The empirical analysis revealed that the EU environmental policy norms, objectives, and strategies had significant influence on those of the Czech Republic. All indicators of policy transfer were clearly detected: Since its EU accession, the country has transposed the EU policy norms and legislation, and implemented them extensively notwithstanding some shortcomings and gaps. This analysis validated the completeness of policy transfer.

What can be the suggestions to move further toward compliance to the EU environmental law? Today's environmental problems have transnational character and require new tools and innovative thinking in addition to more traditional approaches. The key challenge for the EU is to ensure that the best practices in its member states are transmitted and shared in an efficient and effective way to others that may lag behind. When we compared different EU member states and how they adapt to EU pressures regarding environmental policy, we found that some member states, as well as some subnational actors (regions or municipalities), have matured in different degrees on how to handle environmental policy in more efficient manner. In order to provide effective implementation of EU environmental law, the EU has to devise the ways of developing and promoting links with those who have political and administrative power at national, regional, and local (municipal) levels. Besides the top-down relationship between the EU and domestic institutions, this approach also supports the activities in the horizontal dimension in order to reduce the gap between member states or regions. When domestic policy-makers have a clear view on what to do in a specific policy field, this situation foster more proactivity and more political action across levels of politics.

Therefore, in order to ensure effective environmental governance in the horizontal dimension, European Commission introduced the TAIEX-EIR peer-to-peer program in 2017 as a new instrument that facilitates peer learning between experts from member states' environmental authorities. The program helps to share good practice in environmental implementation and governance between member states through expert missions, study visits, and workshops. In this context, experts and employees from an EU member state environmental authority visit an environmental authority in other member states for peer advice and exchange of experience on a specific topic. Also, workshops with environmental authorities from several EU member states are organized with an aim to share good practice in environmental implementation and governance between member states. For instance, a TAIEX-EIR P2P workshop in Galway, Ireland, on February 21–22, 2018, brought together cities participating in the Green Leaf Network, from Spain, Ireland, Belgium, Portugal, and Sweden to work together and share best practices in waste management and the green economy in urban areas. Another conference in Riga, Latvia, on May 27–28, 2019, brought together experts from Estonia, Germany, Latvia, Lithuania, Poland, and Sweden to exchange knowledge and good practices within the Baltic Sea Region countries to advance the integration of sustainability principles and life-cycle cost assessment in public procurement. Furthermore, several reports on best practices are published to inspire environmental authorities.

The focus on the horizontal dimension provides several benefits: First, better strategic relationships can be built between agencies and authorities responsible for implementing and enforcing environmental legislation at national, regional, and local levels. These relationships can be built across thematic areas (such as water and waste management) and across the administrative chain (between policy-makers, inspectors, prosecutors, and auditors). These transnational networks for policy-makers and experts will create strong and stable networks among EU member countries. Secondly, better expert capacity can be developed in agencies and authorities responsible for implementing and enforcing environmental legislation at national, regional, and local levels. Developing knowledge and innovation centers for practitioners on environment can also contribute to the achievement of this goal. These centers can create, host, and coordinate a platform for guidance and knowledge on the environmental acquis through webinars and forums. Finally, dissemination of best practices among the member states can showcase the use of new and emerging techniques and technology, and thus encourage development and spread of new tools, approaches, and methods.

To achieve a sustainable future, we must look beyond the short term. We must also share a common sense of urgency on the challenges ahead. We must have confidence in our ability to set an ambitious agenda for ourselves to protect our environment, and then gear our efforts to delivering it. For all states, regional and municipal administrations, and citizens, it is time to think about the environmental challenges, plan ahead, and make it happen.

Bibliography

Andersen, Svein S. and Eliassen, Kjell A. (eds) (2001) *Making Policy in Europe*, London: Sage Publications.

Bennett C. (1991) 'What is Policy Convergence and What Causes It?', *British Journal of Political Science* (21)2, p. 215–233.

Börzel, T. (1999) 'Towards Convergence in Europe?', *Journal of Common Market Studies*, 39(4), p.573–596.

Börzel, T. (2002) 'Pace-Setting, Foot-Dragging, and Fence-Sitting: Member State Responses to Europeanization', *Journal of Common Market Studies*, 40(2), p. 193–214.

Börzel, T. and Risse, T. (2003) 'Conceptualizing the Domestic Impact of Europe', in Keith Featherstone, Claudio Radaelli (eds.), *The Politics of Europeanisation*, Oxford: Oxford University Press.

Bulmer, S. and Burch, M. (1998) 'Organising for Europe', *Public Administration* 76(4), p. 606–628.

Cowles, M., Caporaso, J. and Risse, T. (eds) (2001) *Transforming Europe: Europeanization and Domestic Change*, Ithaca, NY: Cornell University Press.

DiMaggio, P.J. and Powell, W.W. (1991) 'The Iron Cage Revisited: Institutional Isomorphism and Collective Rationality in Organizational Fields', in Walter W. Powell and Paul J. DiMaggio (eds.), *The New Institutionalism in Organizational Analysis*, Chicago, London: University of Chicago Press, p. 63–83.

Directive 90/313/EEC on the freedom of access to information on the environment, Official Journal of the European Communities, L 158/56.

Directive 2001/42/EC on the assessment of the effects of certain plans and programmes on the environment, Official Journal of the European Communities, L 197/30.

Directive 85/337/EEC on the assessment of the effects of certain public and private projects on the environment, Official Journal of the European Communities, L 175/40.

Directive 2007/2/EC of 14 March 2007 establishing an Infrastructure for Spatial Information in the European Community, Official Journal of the European Union, L 108/1.

Directive 2008/98/EC of the European Parliament and of the Council of 19 November 2008, Official Journal L312/4.

Directive 2003/4/EC of 28 January 2003 on Public Access to Environmental Information, Official Journal of the European Communities, L 41/26.

Directive 91/271/EEC of 21 May 1991 concerning urban waste water treatment, Official Journal of the European Communities, L 135/40.

Directive 75/442/EEC of 15 July 1975 on waste, Official Journal of the European Communities, L 194/39.

Directive 2008/99/EC of 19 November 2008 on the protection of the environment through criminal law, Official Journal of the European Union, L 328/28.

Directive 2018/851/EU of 30 May 2018, Official Journal L 150/109.

Directive 2000/60/EC of 23 October 2000 establishing a framework for Community action in the field of water policy, Official Journal of the European Communities, L 327/1.

Directive 94/62/EC of the European Parliament and of the Council of 20 December 1994, Official Journal L365/10.

Directive 1999/31/EC of the European Parliament and of the Council of 26 April 1999, Official Journal L182/1.

Directive 2018/850/EU of the European Parliament and of the Council of 30 May 2018, Official Journal L150/100.

Directive 2000/76/EC of the European Parliament and of the Council of 4 December 2000, Official Journal L332/91.

Dolowitz, D. and Marsh, D. (1996) 'Who Learns What From Whom: A Review of the Policy Transfer Literature', *Political Studies XLIV*, p.343–357.

EEA (2018) European Waters: Assessment of Status and Pressures 2018, Report No 7/2018.

European Commission (2010) A Strategy for Smart, Sustainable and Inclusive Growth, COM(2010).

European Commission (2015) Closing the Loop - An EU Action Plan for the Circular Economy, COM(2015) 614 final.

European Commission (2015) Assessment of separate collection schemes in the 28 capitals of the EU, Reference: 070201/ENV/2014/691401/SFRA/A2.

European Commission (2019) EU Environmental Implementation Review: Czech Republic, SWD(2019) 119.

European Commission (2019) Second River Basin Management Plans - Czech Republic, SWD(2019) 35 final.

European Council (2001) Conclusions of the Presidency, 15–16 June 2001.

European Court of Auditors (2015) Water Quality in the Danube River Basin, Special Report No.23.

European Environmental Agency (2018) European Waters, Assessment of Status and Pressures, Report No 7/2018.

European Topic Centre on Inland, Coastal and Marine Waters (2015) Results of the Reporting 2011–2013 Under the Drinking Water Directive 98/83/EC.
Featherstone, K. (2003) 'In the Name of Europe', in Featherstone, K. and Radaelli, C. (eds.), *The Politics of Europeanisation*, Oxford: Oxford University Press.
Goldsmith, M. (1993) 'The Europeanisation of Local Government', *Urban Studies* 30(4/5), p. 683–699.
Héritier A. (2001) 'Differential Europe: Administrative Responses to Community Policy' in Maria Green Cowles, James Caporaso and Thomas Risse (eds.), *Transforming Europe: Europeanization and Domestic Change*, Ithaca, NY: Cornell University Press.
Héritier, A., Kerwer, D., Knill, C., Lehmkuhl, D., Teutsch, M. and Douillet, A.C. (2001) *Differential Europe*, Lanham: Rowman and Littlefield.
Héritier, A. and Knill, C. (2001) 'Differential Responses to European Policies: A Comparison' in Maria Green Cowles, James Caporaso and Thomas Risse (eds.), *Transforming Europe: Europeanization and Domestic Change*, Ithaca, NY: Cornell University Press.
Holzinger, K., Knill, C., Sommerer, T. (2008) Environmental Policy Convergence: The Impact of International Harmonization, Transnational Communication, and Regulatory Competition, *International Organization*, 62(4), p. 553–587.
Hrabankova, A. and Martinkova, M. (2010) 'Implementation of EC Nitrate Directive in the Czech Republic', 38th IAD Conference, Dresden, Germany.
Hyrslova, J. and Hajek, M. (2005) 'Environmental Management Accounting in the Framework of EMAS II in the Czech Republic', in Rikhardsson, P., Bennett, M., Bouma, J., Schaltegger, S. (eds.), *Implementing Environmental Management Accounting: Status and Challenges*, The Netherlands: Springer, p. 279–295.
Jacoby, W. (1999) 'Tutors and Pupils: International Organisations, Central European Elites, and Western Models', *Governance*, 14(2), p. 169–200.
Knill, C. and Lehmkuhl, D. (1999) 'How Europe Matters: Different Mechanisms of Europeanization', *European Integration Online Papers* 3(7).
Knill, C. and Lenschow, A. (2001) 'Adjusting to EU Environmental Policy: Change and Persistance of Domestic Administrations', in Cowles, M., Caporaso, J., Risse, T. (eds.), *Transforming Europe*, Cornell University Press.
Ladrech, R. (1994) 'Europeanization of Domestic Politics and Institutions: The Case of France', *Journal of Common Market Studies*, 32(1), p. 69–88.
Margaux, L., Sauquet, A., and Subervie J. (2019) 'Providing technical assistance to peer networks to reduce pesticide use in Europe: Evidence from the French Ecophyto plan', https://hal.archives-ouvertes.fr/hal-02190979v2
Ministry of Agriculture (2012) Report on Water Management, Prague.
Ministry of Agriculture (2016) Report on Water Management, Prague.
Ministry of Agriculture (2017) Report on Water Management, Prague.
Ministry of Environment (1996) Report on the Environment in the Czech Republic, Prague.
Ministry of Environment (2012) State Environmental Policy of the Czech Republic, Prague.

Ministry of Environment (2014) Report on the Environment in the Czech Republic, Prague.

Ministry of Environment (2014) Waste Management Plan of the Czech Republic, Prague.

Ministry of Environment (2015) Strategy on Adaptation to Climate Change in the Czech Republic, Government Resolution No.861, Prague.

Ministry of Environment (2016) Report on the Environment in the Czech Republic, Prague.

Moldan, B. and Hak, T. (2007) 'Environment in the Czech Republic: A Positive and Rapid Change', *Environmental Science and Technology*, 15 January 2007, p.359–362.

Nicolaides, P. (1999) *Enlargement of the EU and Effective Implementation of Community Rules: An Integration-Based Approach*, Maastricht: European Institute for Public Administration Working Paper.

OECD (2018) OECD Environmental Performance Reviews: Czech Republic 2018, OECD Publishing, Paris.

Office of the Government of the Czech Republic (2017) Strategic Framework Czech Republic 2030, online source: www.cr2030.cz, printed version: ISBN 978-80-7440-181-7.

Olsen, J. P. (1996) 'Europeanization and Nation-State Dynamics', in Gustavsson S. and Lewin L. (eds.), *The Future of the Nation-State*, London: Routledge.

Radaelli, C. (2000) 'Policy Transfer in the European Union: Institutional Isomorphism as a Source of Legitimacy', *Governance* 13(1), p. 25–43.

Radaelli, C. (2003) 'The Europeanization of Public Policy', in Keith Featherstone, Claudio Radaelli (eds.), *The Politics of Europeanisation*, Oxford: Oxford University Press.

Reiche, D. (2006) 'Renewable Energies in the EU-Accession States', *Energy Policy* 34, p.365–375.

Rometsch, D. and Wessels, W. (eds) (1996) *The EU and Member States*, Manchester: Manchester University Press.

Tanil, G. (2012) *Europeanization, Integration and Identity*, New York: Routledge.

Index

About the Author

Writer, researcher, and academic, **Gamze Tanil** is the author of three books about European politics, public policies, and international relations. She has a PhD in politics and international relations from University of Liverpool. With over a decade teaching subjects as European history, diplomacy, politics, and sustainable development, she has a unique voice that comes through her lectures on the importance we place on the environment and policy. She has travelled extensively around Central and Eastern Europe, learning about the history and politics of the region. When not absorbed in reading about environmental issues and policy proposals for the sake of the world's common future, she loves watching films and documentary series on history and nature, enjoys traveling to new destinations to discover new cultures, lifestyles, art, and food, and otherwise spends far too much time on the computer.

www.ingramcontent.com/pod-product-compliance
Ingram Content Group UK Ltd.
Pitfield, Milton Keynes, MK11 3LW, UK
UKHW020842290926
15470UKWH00004B/1500

9 781793 6338